GREENHOUSE GASES AND THEIR IMPACT

GREENHOUSE GASES AND THEIR IMPACT

BERNARD F. SHULK
EDITOR

Nova Science Publishers, Inc.
New York

For permission to use material from this book please contact us:
Telephone 631-231-7269; Fax 631-231-8175
Web Site: http://www.novapublishers.com

Library of Congress Cataloging-in-Publication Data

Greenhouse gases / Bernard F. Shulk (editor).
p. cm.
Includes bibliographical references and index.
ISBN 13 978 1-60021-392-2
ISBN 10 1-60021-392-8
1. Greenhouse gas mitigation--United States. 2. Greenhouse gases--United States. I. Shulk, Bernard F.
TD885.5.G73G77 2006
363.738'74--dc22 2006024581

Published by Nova Science Publishers, Inc. ✦New York

CONTENTS

PREFACE

As the U.S. population increases, the need for energy sources such as electricity, petroleum and industrial processes and the resulting emissions are growing at an alarming rate. Emissions that increase the rate of global warming include carbon dioxide, methane, nitrous oxide and other greenhouse gases. This book examines current challenges encountered in attempting to reduce the consumption of fossil fuels as well as developing alternative methods of producing energy.

Chapter 1 - This report—the thirteenth annual report—presents the Energy Information Administration's latest estimates of emissions for carbon dioxide, methane, nitrous oxide, and other greenhouse gases. These estimates are based on activity data and applied emissions factors and not on measured or metered emissions.

Chapter 2 - This report reviews U.S. emissions of greenhouse gases in the contexts both of domestic policy and of international obligations and proposals. On October 15, 1992, the United States ratified the United Nations Framework Convention on Climate Change (UNFCCC), which entered into force on March 21, 1994. This committed the United States to "national policies" to limit "its anthropogenic emissions of greenhouse gases," with a voluntary goal of returning "emissions of carbon dioxide [CO_2] and other greenhouse gases [methane (CH_4), nitrous oxide (N_2O),hydrofluorocarbons (HFCs), perfluorocarbons (PFCs), and sulfurhexafluoride (SF_6)]" at the "end of the decade" to "their 1990 levels."

Subsequently, in the 1997 Kyoto Protocol to the UNFCCC, the United States participated in negotiations that ended with agreement on further reductions that could become legally binding. The United States signed the Kyoto Protocol in 1998, but President Clinton did not send it to the Senate for advice and consent. President Bush has said that he rejects the Protocol, and former U.S. Environmental Protection Agency Administrator Christine Todd Whitman told reporters that the Administration would not be pursuing the UNFCCC commitment either. Instead, President Bush has proposed to shift the nation's climate change program from a goal of reducing emissions per se to a goal of reducing energy intensity — the amount of greenhouse gases emitted per unit of economic productivity. Under the proposal, the intensity, which has been declining for a number of years, would decline 18% between 2002 and 2012, as opposed to a 14% projected "business as usual" decline.

Meanwhile, the UNFCCC "end of the decade" deadline has passed and U.S. greenhouse gas emissions continue on an upward trend, though with dips in 1991and in 2001, attributed mostly to economic slowdowns. Based on historical data, 2001 emissions were about 13% in excess of the UNFCCC goal. Overall, from 1990 to 2001, U.S. greenhouse gas emissions (weighted by global warming potential) have increased an average of about 1.1% per year. Projections suggest that U.S. emissions will continue to rise for at least the next decade. Reversing the upward trend in greenhouse gas emissions would represent an extraordinary technical and political challenge to U.S. energy and environmental policy.

Chapter 3 - The possibility that human activities are releasing gases, including carbon dioxide (CO2), at rates that could affect global climate has resulted in proposals for national pro-grams to curtail emissions. An international framework for specific reductions in greenhouse gases was negotiated at a meeting in Kyoto in December 1997. Concern about costs has encouraged consideration of CO2 reduction proposals that employ market-based mechanisms. The passage in 1990 of a trade-able allowance system for sulfur dioxide (SO2) control in the United States to reduce acid rain provides a precedent for such mechanisms.

The two mechanisms receiving the most attention are a tradeable permit program (similar to the acid rain program) and carbon taxes. Proposed CO2 reduction schemes present large uncertainties in terms of the perceived reduction needs and the potential costs of achieving those reductions. Tradeable permit programs would reduce CO2 emissions to a specific level with the control cost han-dled efficiently, but not ata specific cost level. Carbon taxes would effectively cap marginal control costs at the specific tax level, but the precise level of CO2 reduction achieved would be less certain. Hence, a major policy question is whether one is more concerned about the possible cost of the program and therefore willing to accept some uncertainty about emission reduction in order to have some limits on costs (i.e., carbon taxes) or whether one is more concerned about achieving a specific emission reduction level with costs handled efficiently, but not capped (i.e., tradeable permits).

The specific effects of both a carbon tax and tradeable permit program would depend on the specific levy (carbon tax) or allocation scheme (tradeable permit) chosen, the scope of the program, the timing of the reductions, and the recycling of any revenues.

In addition, many tradeable permit proposals include provisions allowing countries to accumulate permits by reducing emissions in other countries. Two such schemes, joint implementation and the Clean Development Mechanism (CDM), were approved at the Kyoto conference in December 1997.

The climate change issue and CO2 control raise numerous equity issues. In one sense, climate change is a concern about intergenerational equity — the well-being of the current generation versus generations to come. On a global level, the issue also involves the North-South debate. At the domes-tic level, equity questions include the regional distribution of costs under a tradeable permit or carbon tax scheme. For example, an important impact of either a carbon tax based on the carbon content of fossil fuels or a tradeable permit program would be the pressure for fuel shifts away from coal and toward gas. Regions such as fast-growing areas in need of more energy and owners of "all electric" homes, among others, would likely be disproportionately hit by a CO2 control scheme. In addition, people may be affected differently according to income class. These issues, however, have not been sufficiently analyzed at the current time to be sure of how various sectors would be affected.

In: Greenhouse Gases and Their Impact
Editor: Bernard F. Shulk, pp. 1-142
ISBN: 978-1-60021-392-2

Chapter 1

U.S. EMISSIONS OF GREENHOUSE GASES: BACKGROUND AND CONTEXT

Paul McArdle

SUMMARY

This report, *Emissions of Greenhouse Gases in the United States 2004,* was prepared under the general direction of John Conti, Director of the Office of Integrated Analysis and Forecasting, Energy Information Administration. General questions concerning the content of this report may be directed to the National Energy Information Center at 202/586-8800.

Specific technical information concerning the content of the report may be obtained from Perry Lindstrom at 202/586-0934 (e-mail, perry.lindstrom@eia.doe.gov). Without the assistance of Science Applications International Corporation this report would not have been possible. In particular we would like to thank Nancy Checklick, Keith Forbes, Michael Mondshine, and Brennen Walsh.

The authors would also like to express their gratitude to all the people who provided information or comments on this report. In particular, Mark Schipper of EIA's Office of Energy Markets and End Use provided survey data and analysis from the 2002 Manufacturing Energy Consumption Survey; and Stan Kaplan of EIA's Office of Coal, Nuclear, Electric and Alternate Fuels provided electric power sector data and analysis. We also thank Leif Hockstad of the U.S. Environmental Protection Agency for expediting valuable comments on the draft report.

Title XVI, Section 1605(a) of the Energy Policy Act of 1992 (enacted October 24, 1992) provides:

> *Not later than one year after the date of the enactment of this Act, the Secretary, through the Energy Information Administration, shall develop, based on data available to, and obtained by, the Energy Information Administration, an inventory of the national aggregate emissions of each greenhouse gas for each calendar year of the baseline period of 1987 through 1990. The Administrator of the Energy Information Administration shall annually update and analyze such inventory using available data. This subsection does not provide any new data collection authority.*

The first report in this series, *Emissions of Greenhouse Gases 19851990,* was published in September 1993. This report—the thirteenth annual report—presents the Energy Information Administration's latest estimates of emissions for carbon dioxide, methane, nitrous oxide, and other greenhouse gases. These estimates are based on activity data and applied emissions factors and not on measured or metered emissions.

For this report, data on coal and natural gas consumption and electricity sales and losses by sector were obtained from the Energy Information Administration's (EIA's) September 2005 *Monthly Energy Review.* Additional detailed information on petroleum consumption was obtained from unpublished material in support of EIA's *Annual Energy Review 2004.* Electric power sector emissions were obtained from data underlying EIA's *Electric Power Annual.* In keeping with current international practice, this report presents data on greenhouse gas emissions in million metric tons carbon dioxide equivalent. The data can be converted to carbon equivalent units by multiplying times 12/44.

INTRODUCTION

Overview

U.S. Anthropogenic Emissions of Greenhouse Gases, 1990-2004

	Carbon Dioxide Equivalent
Estimated 2004 Emissions (Million Metric Tons)	7,122.1
Change Compared to 2003 (Million Metric Tons)	138.8
Change from 2003 *(Percent)*	*2.0%*
Change Compared to 1990 (Million Metric Tons)	973.3
Change from 1990 *(Percent)*	*15.8%*
Average Annual Increase, 1990-2004 *(Percent)*	*1.1%*

U.S. emissions of greenhouse gases in 2004 totaled 7,122.1 million metric tons carbon dioxide equivalent (MMTCO$_2$e), 2.0 percent more than in 2003 (6,983.2 MMTCO$_2$e). The 2004 increase in total greenhouse gas emissions is attributable primarily to a 1.7percent increase in emissions of carbon dioxide, along with increases in emissions of nitrous oxide (5.5 percent) and methane (0.9 percent). Emissions of engineered gases— hydrofluorocarbons (HFCs), perfluorocarbons (PFCs), and sulfur hexafluoride (SF_6)—also increased, by 9.6 percent. The U.S. economy grew by 4.2 percent in 2004, the highest rate of growth since 1999. Consequently, U.S. greenhouse gas intensity (greenhouse gas emissions per unit of real economic output) was 2.1 percent lower in 2004 than in 2003. From 1990 to 2004, U.S. greenhouse gas intensity declined by 23 percent, or by an average of 1.9 percent per year.

U.S. greenhouse gas emissions in 2004 were 16 percent higher than the 1990 emissions level of 6,148.8 MMTCO$_2$e—an average annual increase of 1.1 percent over the period. Since 1990, U.S. emissions have increased more slowly than the average annual growth in population (1.2 percent), primary energy consumption (1.2 percent), electric power generation (1.9 percent), or gross domestic product (3.0 percent). While the annual growth rate in carbon dioxide emissions since 1990 (1.3 percent) has closely tracked annual growth in population and energy consumption, the average annual rate of growth in total greenhouse gas emissions has been lower (1.1 percent) because of reductions in methane emissions and low growth in nitrous oxide emissions (0.7 percent) since 1990.

Table ES1 shows trends in emissions of the principal greenhouse gases, measured in million metric tons of native gas. In Table ES2, emissions trends are shown in carbon dioxide equivalents, which are derived by multiplying the volumes of native gas by the global warming potential (GWP) of each gas. The GWPs provide a measure of marginal radiative efficiency for each of the native gases. The GWP concept, developed by the Intergovernmental Panel on Climate Change (IPCC), provides a comparative measure of the impacts of added units of different greenhouse gases on global warming relative to the impact of carbon dioxide.[1] This allows for the normalization of different greenhouse gases, which have different warming impacts, into a common unit of measure—carbon dioxide equivalents—which can be aggregated across gases (see box on page xi).

Table ES1. Summary of Estimated U.S. Emissions of Greenhouse Gases, 1990 and 1996-2004 (Million Metric Tons of Gas)

Gas	1990	1996	1997	1998	1999	2000	2001	2002	2003	P2004
Carbon Dioxide.	5,002.3	5,499.7	5,563.0	5,598.1	5,677.9	5,845.5	5,785.5	5,808.5	5,871.8	5,973.0
Methane	31.4	29.4	29.4	28.4	27.9	27.8	27.2	27.2	27.6	27.8
Nitrous Oxide	1.1	1.2	1.2	1.2	1.2	1.2	1.1	1.1	1.1	1.2
HFCs, PFCs, and SF_6 . .	M	M	M	M	M	M	M	M	M	M

M = mixture of gases. These gases cannot be summed in native units. See Table ES2 for estimated totals in carbon dioxide equivalent. P = preliminary data.

Note: Data in this table are revised from the data contained in the previous EIA report, *Emissions of Greenhouse Gases in the United States 2003,* DOE/EIA0573(2003) (Washington, DC, December 2004).

Source: Estimates presented in this report.

Table ES2. U.S. Emissions of Greenhouse Gases, Based on Global Warming Potential, 1990 and 1996-2004 (Million Metric Tons Carbon Dioxide Equivalent)

Gas	1990	1996	1997	1998	1999	2000	2001	2002	2003	P2004
Carbon Dioxide	5,002.3	5,499.7	5,563.0	5,598.1	5,677.9	5,845.5	5,785.5	5,808.5	5,871.8	5,973.0
Methane.	721.4	675.8	675.2	654.2	642.2	639.8	625.8	626.2	633.9	639.5
Nitrous Oxide.	337.0	358.0	349.1	348.8	347.1	343.5	338.8	335.1	335.2	353.7
HFCs, PFCs, and SF_6 . .	88.1	114.3	122.0	137.7	137.4	142.1	133.9	143.1	142.4	155.9
Total	**6,148.8**	**6,647.7**	**6,709.3**	**6,738.8**	**6,804.7**	**6,970.8**	**6,884.1**	**6,912.9**	**6,983.2**	**7,122.1**

P = preliminary data.

Note: Data in this table are revised from the data contained in the previous EIA report, *Emissions of Greenhouse Gases in the United States 2003,* DOE/EIA0573(2003) (Washington, DC, December 2004).

Sources: Emissions: Estimates presented in this report. **Global Warming Potentials:** Intergovernmental Panel on Climate Change, *Climate Change 2001: The Scientific Basis* (Cambridge, UK: Cambridge University Press, 2001), pp. 38 and 388389.

In 2001, the IPCC Working Group I released its Third Assessment Report, *Climate Change 2001: The Scientific Basis.* [2] Among other things, the Third Assessment Report updated a number of the GWP estimates that appeared in the IPCC's Second Assessment Report.[3] The GWPs published in the Third Assessment Report were used for the calculation of carbon dioxide equivalent emissions for this report. Generally, the level of total U.S. carbon dioxide equivalent emissions is 0.7 percent higher when the GWPs from the Third Assessment Report are used; however, the trends in growth of greenhouse gas emissions are similar for both sets of GWP values.

During 2004, 82.4 percent of total U.S. greenhouse gas emissions consisted of carbon dioxide from the combustion of fossil fuels such as coal, petroleum, and natural gas (after adjustments for U.S. Territories and international bunker fuels). U.S. emissions trends are driven largely by trends in fossil energy consumption. In recent years, national energy consumption, like emissions, has grown relatively slowly, with yeartoyear deviations from trend growth caused by weatherrelated phenomena, fluctuations in business cycles, changes in the fuel mix for electric power generation, and developments in domestic and international energy markets.

Other 2004 U.S. greenhouse gas emissions (Figure ES1) include carbon dioxide from noncombustion sources (1.5 percent of total U.S. greenhouse gas emissions), methane (9.0 percent), nitrous oxide (5.0 percent), and other gases (2.2 percent). Methane and nitrous oxide emissions are caused by the biological decomposition of various waste streams and fertilizer; fugitive emissions from chemical processes; fossil fuel production, transmission, and combustion; and many smaller sources. The other gases include HFCs, used primarily as refrigerants; PFCs, released as fugitive emissions from aluminum smelting and used in semiconductor manufacture; and SF_6, used as an insulator in utilityscale electrical equipment.

This report, in accordance with Section 1605(a) of the Energy Policy Act of 1992, provides estimates of U.S. emissions of greenhouse gases. The estimates are based on activity data and applied emissions factors, not on measured or metered emissions.

Carbon Dioxide

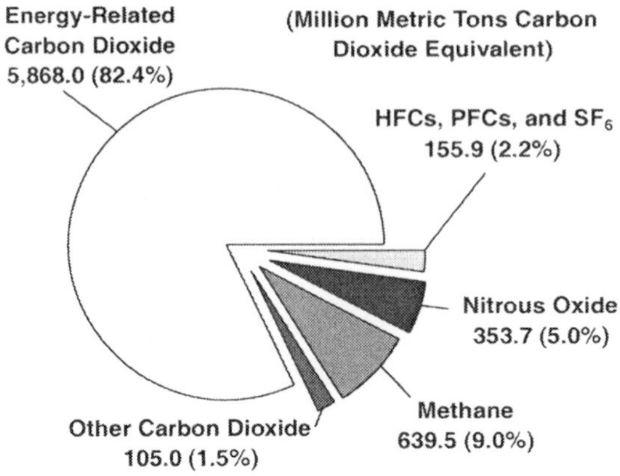

Source: Tables ES2 and ES3.

Figure ES1. U.S. Greenhouse Gas Emissions by Gas, 2004

The preliminary estimate of U.S. carbon dioxide emissions from both energy consumption and industrial processes in 2004 is 5,973.0 million metric tons (MMT), which is 1.7 percent higher than in 2003 (5,871.8 MMT) and accounts for 83.9 percent of total U.S. greenhouse gas emissions (see Table ES3 for a breakdown of U.S. carbon dioxide emissions by source).

Units for Measuring Greenhouse Gases

Emissions data are reported here in metric units, as favored by the international scientific community. Metric tons are relatively intuitive for users of English measurement units, because 1 metric ton is only about 10 percent heavier than 1 English short ton.

Table ES1 shows emissions of greenhouse gases in terms of the full molecular weights of the native gases. In Table ES2, and subsequently throughout this report, emissions of carbon dioxide and other greenhouse gases are given in carbon dioxide equivalents. In the case of carbon dioxide, emissions denominated in the molecular weight of the gas or in carbon dioxide equivalents are the same. Carbon dioxide equivalent data can be converted to carbon equivalents by multiplying by 12/44.

Table ES3. U.S. Carbon Dioxide Emissions from Energy and Industry, 1990 and 1996-2004 (Million Metric Tons Carbon Dioxide)

Fuel Type or Process	1990	1996	1997	1998	1999	2000	2001	2002	2003	P2004
Energy Use by Sector										
Residential	953.7	1,090.2	1,082.0	1,088.5	1,110.9	1,171.8	1,168.3	1,190.4	1,213.2	1,212.0
Commercial	780.7	872.1	916.1	934.1	947.3	1,005.9	1,026.2	1,019.8	1,021.1	1,024.2
Industrial	1,692.2	1,788.8	1,804.4	1,791.2	1,779.1	1,780.3	1,698.4	1,671.0	1,685.6	1,730.2
Transportation	1,569.9	1,705.3	1,722.7	1,757.9	1,806.0	1,844.2	1,835.5	1,864.9	1,875.7	1,933.7
Energy Subtotal	**4,996.6**	**5,456.5**	**5,525.2**	**5,571.6**	**5,643.2**	**5,802.3**	**5,728.4**	**5,746.0**	**5,795.5**	**5,899.9**
Nonfuel Use Emissions[a]	**98.1**	**105.6**	**110.5**	**118.6**	**124.1**	**110.0**	**104.9**	**107.0**	**105.2**	**114.3**
Nonfuel Use Sequestration	**251.2**	**291.2**	**303.4**	**314.7**	**325.8**	**308.1**	**293.9**	**298.3**	**294.0**	**316.6**
Adjustments to Energy										
U.S. Territories (+)	31.1	37.7	39.0	41.3	40.7	42.2	53.6	52.2	58.4	61.7
Military Bunker Fuels (-)	13.6	8.9	9.6	10.0	9.8	7.8	8.2	8.1	9.2	9.2
International Bunker Fuels (-)	100.1	93.3	100.1	104.9	97.4	93.5	89.6	81.2	75.0	84.4
Bunker Fuels Subtotal (-)	*113.7*	*102.2*	*109.7*	*114.9*	*107.2*	*101.3*	*97.8*	*89.3*	*84.2*	*93.6*
Total Energy Adjustments	**-82.6**	**-64.5**	**-70.7**	**-73.6**	**-66.5**	**-59.1**	**-44.2**	**-37.1**	**-25.9**	**-31.9**
Adjusted Energy Subtotal	**4,914.0**	**5,392.0**	**5,454.6**	**5,498.0**	**5,576.8**	**5,743.1**	**5,684.2**	**5,709.0**	**5,769.6**	**5,868.0**
Other Sources										
Natural Gas Flaring	9.1	16.5	15.5	6.2	6.7	5.5	5.9	6.0	5.9	5.9
Carbon Dioxide in Natural Gas	14.0	17.8	18.0	18.0	17.8	18.2	18.6	17.9	18.0	17.8
Cement Production	33.3	37.2	38.4	39.3	40.1	41.3	41.5	43.0	43.2	44.8
Other Industrial	26.8	29.0	29.2	29.7	29.3	29.4	27.4	26.4	27.6	28.7
Waste Combustion	5.1	7.3	7.3	6.9	7.2	7.9	8.0	6.2	7.5	7.8
Total Other Sources	**88.3**	**107.7**	**108.5**	**100.1**	**101.2**	**102.3**	**101.3**	**99.5**	**102.2**	**105.0**
Total	**5,002.3**	**5,499.7**	**5,563.0**	**5,598.1**	**5,677.9**	**5,845.5**	**5,785.5**	**5,808.5**	**5,871.8**	**5,973.0**

[a]Emissions from nonfuel use are included in the sectoral totals above.

P = preliminary data.

Notes: Data in this table are revised from the data contained in the previous EIA report, *Emissions of Greenhouse Gases in the United States 2003,* DOE/EIA0573(2003) (Washington, DC, December 2004). Totals may not equal sum of components due to independent rounding. Adjusted energy total includes U.S. Territories.

Sources: EIA estimates presented in this report.

Emissions of other greenhouse gases (such as methane) can also be measured in carbon dioxide equivalent units by multiplying their emissions (in metric tons) by their global warming potentials (GWPs).

Carbon dioxide equivalents are the amount of carbon dioxide by weight emitted into the atmosphere that would produce the same estimated radiative forcing as a given weight of another radiatively active gas. Carbon dioxide equivalents are computed by multiplying the weight of the gas being measured (for example, methane) by its estimated GWP (which is 23 for methane). GWPs are discussed in Chapter 1 and summarized in Table 4 on page 13.

U.S. carbon dioxide emissions have grown by an average of 1.3 percent annually since 1990. Although shortterm changes in carbon dioxide emissions can result from temporary variations in weather, power generation fuel mixes, and the economy, growth in carbon dioxide emissions in the longer term results largely from population and incomedriven increases in energy use, as well as consumer choices of energyusing equipment. The "carbon intensity" of energy use (carbon dioxide emissions per unit of energy consumed) can also influence the trend of growth in energyrelated carbon dioxide emissions.

Figure ES2 shows recent trends in some common indexes used to measure the carbon intensity of the U.S. economy. Carbon dioxide emissions per unit of gross domestic product (GDP) have continued to fall relative to 1990; by 2004, this measure was 21 percent lower than in 1990. Carbon dioxide emissions per capita, however, were 1.2 percent above 1990 levels in 2004. Population growth and other factors resulted in increased aggregate carbon dioxide emissions per year from 1990 through 2004 (a total increase of 19 percent). Carbon dioxide emissions per unit of net electricity generation in 2004 were 1.1 percent lower than in 2003.

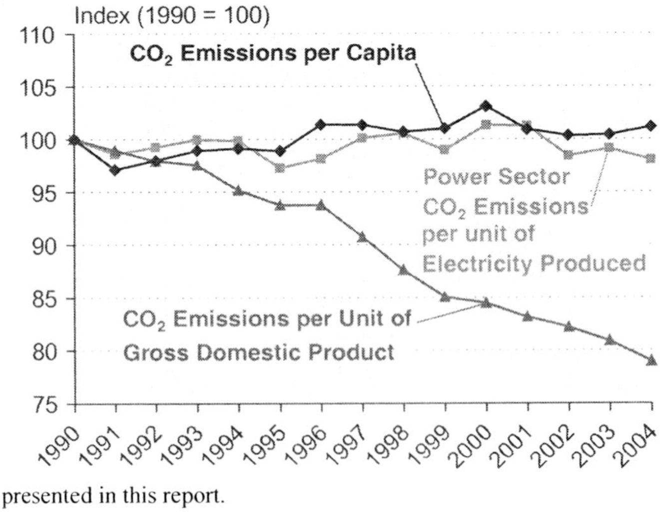

Sources: Estimates presented in this report.

Figure ES2. Carbon Dioxide Emissions Intensity of U.S. Gross Domestic Product, Population, and Electricity Production, 1990-2004

Energy Consumption

The consumption of energy in the form of fossil fuel combustion is the largest single contributor to anthropogenic greenhouse gas emissions in the United States and the world. Of total 2004 U.S. carbon dioxide emissions (adjusting for U.S. Territories and bunker fuels), 98 percent, or 5,868.0 MMT, resulted from the combustion of fossil fuels. This figure represents an increase of 1.7 percent from 2003 levels.

The Energy Information Administration (EIA) divides energy consumption into four general enduse categories: residential, commercial, industrial, and transportation. Emissions from electricity generators, which provide electricity to the enduse sectors, are allocated in proportion to the electricity consumed in, and losses allocated to, each sector. Figure ES3 illustrates trends in carbon dioxide emissions by energy consumption sector. Average annual growth rates in carbon dioxide emissions by sector during the 1990-2004 period were 2.0 percent for the commercial sector, 1.7 percent for the residential sector, 1.5 percent for the transportation sector, and 0.2 percent for the industrial sector.

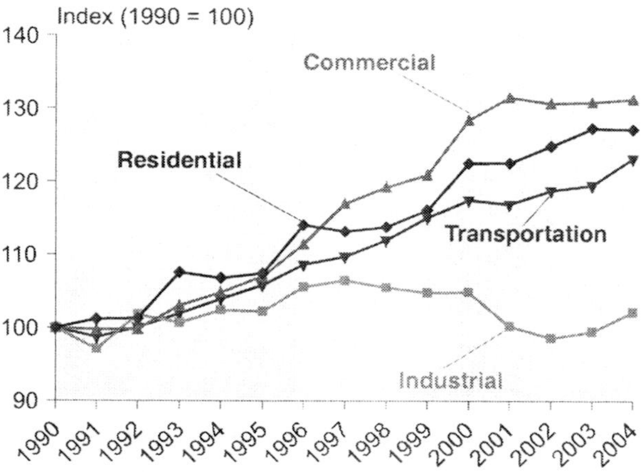

Note: Sectoral emissions include both direct emissions and emissions attributable to purchased electricity.
Sources: Estimates presented in this report.

Figure ES3. U.S. Carbon Dioxide Emissions by Sector, 1990-2004

Carbon dioxide emissions from the transportation sector are the largest source of energyrelated carbon dioxide emissions. At 1,933.7 MMT, the transportation sector accounted for 33 percent of total U.S. energyrelated carbon dioxide emissions in 2004. Transportation sector emissions increased by 3.1 percent in 2004 relative to the 2003 level of 1,875.7 MMT. Almost all (98 percent) of transportation sector carbon dioxide emissions result from the consumption of petroleum products: motor gasoline, 1,162.6 MMT (60 percent of total transportation sector emissions in 2004); middle distillates (diesel fuel), 428.2 MMT (22 percent); jet fuel, 237.4 MMT (12 percent); and residual oil (heavy fuel oil, largely for maritime use), 54.6 MMT (2.8 percent). The growth in transportationrelated carbon dioxide emissions in 2004 included increases in emissions from the use of motor gasoline (21.2 MMT), diesel fuel (17.9 MMT), residual fuel oil (10.0 MMT), and jet fuel (8.2 MMT).

Industrial sector carbon dioxide emissions, at 1,730.2 MMT, accounted for 29 percent of total U.S. energyrelated carbon dioxide emissions in 2004. The 2004 emissions level represents a 2.6percent increase over 2003 emissions of 1,685.6 MMT, the largest since a 3.0percent increase from 1995 to 1996. The 2004 increase in industrial emissions resulted from strong growth in industrial production, up 4.1 percent from 2003. In terms of fuel shares, electricity consumption was responsible for 38 percent of total industrial sector carbon dioxide emissions (660.9 MMT), natural gas for 26 percent (441.9 MMT), petroleum for 25 percent (440.6 MMT), and coal for 10 percent (181.0 MMT). Carbon dioxide emissions attributable to industrial sector energy consumption, while fluctuating from year to year, have increased by an average of 0.2 percent per year since 1990. As a result, total energyrelated industrial emissions in 2004 were 2.2 percent (38.0 MMT) higher than in 1990.

At 1,212.0 MMT, residential carbon dioxide emissions represented 21 percent of U.S. energyrelated carbon dioxide emissions in 2004. The 2004 residential emissions were 0.1 percent lower than the 2003 level of 1,213.2 MMT. A warmer winter in 2004, relative to 2003, was a contributor to the 2004 decrease in residential sector emissions. Additionally, while the winter was warmer than 2003, the summer was cooler, leading to a drop of about 4 percent in both heating and cooling degreedays.[4] The residential sector's prorated share of electric power sector carbon dioxide emissions, 837.3 MMT, accounted for more than twothirds of all emissions in the residential sector.[5] Natural gas accounted for 22 percent (265.5 MMT) and petroleum (mainly distillate fuel oil) represented 8.9 percent (108.0 MMT). Since 1990, when residential sector carbon dioxide emission totaled 953.7 MMT, the growth in residential carbon dioxide emissions has averaged 1.7 percent per year.

Commercial sector carbon dioxide emissions accounted for about 17 percent of total energyrelated carbon dioxide emissions in 2004, at 1,024.2 MMT, of which 78 percent (795.4 MMT) was the sector's prorated share of electricityrelated emissions. Natural gas contributed 16 percent (162.7 MMT) and petroleum 5.7 percent (57.9 MMT). Commercial sector carbon dioxide emissions increased by 0.3 percent from the 2003 level of 1,021.1 MMT. Since 1990, carbon dioxide emissions in the commercial sector have increased on average by 2.0 percent per year, the largest growth of any enduse sector. Commercial sector carbon dioxide emissions have risen by 243.4 MMT since 1990, accounting for 27 percent of the total increase in U.S. energyrelated carbon dioxide emissions.

Carbon dioxide emissions from the U.S. electric power sector increased by 0.9 percent (19.7 MMT), from 2,278.8 MMT in 2003 to 2,298.6 MMT in 2004. Carbon dioxide emissions from the electric power sector have grown by 27 percent since 1990, while total carbon dioxide emissions from all energyrelated sources have grown by 18 percent. Carbon dioxide emissions from the electric power sector represented 39 percent of total U.S. energyrelated carbon dioxide emissions in 2004; however, as noted above, in calculating emissions from the enduse sectors EIA distributes electric power sector emissions to the four sectors in proportion to their respective shares of total electricity purchases. Therefore, electric power emissions are already included in the sectoral totals. By fuel, emissions from naturalgasfired generation increased by 6.6 percent, emissions from coalfired generation increased by 0.1 percent, and emissions from petroleumfired generation increased by 0.3 percent in 2004 from their 2003 levels.

Nonfuel Uses of Energy Inputs

Nonfuel uses of fossil fuels, principally petroleum, both emit carbon dioxide and sequester carbon over their life cycles. In 2004, nonfuel uses of fossil fuels resulted in emissions of 114.3 MMT carbon dioxide (Table ES3), an 8.6percent increase from the 2003 level of 105.2 MMT. Emissions from nonfuel uses of energy fuels are included in the unadjusted energy consumption subtotals in Table ES3.

On the sequestration side of the ledger, nonfuel uses of fossil fuels also resulted in carbon sequestration equal to 316.6 $MMTCO_2e$ in 2004, a 7.7percent increase from the 2003 level of 294.0 $MMTCO_2e$.[6] The major fossil fuel products that emit and sequester carbon include liquefied petroleum gas (LPG) and feedstocks for plastics and other petrochemicals. Asphalt and road oils are a major source of sequestration, but they do not emit carbon dioxide. It is estimated that, of the amount of carbon sequestered in the form of plastic, about 11.0 $MMTCO_2e$ was emitted as carbon dioxide from the burning of the plastic components of municipal solid waste to produce electricity in 2004. The 2003 estimate of 18.8 $MMTCO_2e$ is used in this report as an estimate for total 2004 emissions of carbon dioxide from the burning of wastes. The U.S. Environmental Protection Agency (EPA) estimates total emissions from waste burning, and its 2004 value was not available at the time this report was published.

Adjustments to Energy Consumption

Total U.S. carbon dioxide emissions and the estimates of energy consumption on which they are based correspond to EIA's coverage of energy consumption, including the 50 States and the District of Columbia. Under the United Nations Framework Convention on Climate Change (UNFCCC), however, the United States is also responsible for counting emissions emanating from its Territories, and their emissions are added to the U.S. total. Conversely, because the IPCC definition of energy consumption excludes international bunker fuels from the statistics of all countries, emissions from international bunker fuels are subtracted from the U.S. total. Military bunker fuels are also subtracted, because they are also excluded by the IPCC from national emissions totals. On net, these adjustments resulted in the subtraction of 31.9 MMT from total U.S. carbon dioxide emissions related to energy consumption (5,899.9 MMT), resulting in an adjusted total of 5,868.0 MMT for energyrelated carbon dioxide emissions in 2004 (Table ES3).

Energyrelated carbon dioxide emissions for the U.S. Territories are added as an adjustment, in keeping with IPCC guidelines for national emissions inventories. The Territories included are Puerto Rico, the U.S. Virgin Islands, American Samoa, Guam, the U.S. Pacific Islands, and Wake Island. Most of these emissions are from petroleum products; however, Puerto Rico and the Virgin Islands consume coal in addition to petroleum products. Total energyrelated carbon dioxide emissions from the U.S. Territories in 2004 are estimated at 61.7 MMT (Table ES3).

For 2003, the carbon dioxide emissions estimate for military bunker fuels was 9.2 MMT.[7] In 2004, approximately 93.6 MMT carbon dioxide was emitted in total from international bunker fuels, including 84.4 MMT attributed to civilian consumption of bunker fuels. In Table ES3, that amount has been subtracted from the U.S. total. Just over onehalf of the carbon dioxide emissions associated with international bunker fuels comes from the combustion of jet fuels; residual and distillate fuels account for the other half, with most of that coming from residual fuel.

Other Carbon Dioxide Emissions

In addition to carbon dioxide emission from fossil fuel combustion and use, a total of 105.0 MMT was emitted from other sources in 2004 (Table ES3). Cement manufacture (44.8 MMT) and industrial sources (28.7 MMT) accounted for nearly threefourths of the total carbon dioxide emissions from other sources. Energy sector components in the other emissions category included the stripping of carbon dioxide from natural gas (17.8 MMT) and natural gas flaring (5.9 MMT). An additional 7.8 MMT carbon dioxide is estimated to have been released from the burning of wastes other than municipal solid waste in the electric power sector.

Methane

U.S. anthropogenic methane emissions totaled 639.5 $MMTCO_2e$[8] (27.8 million metric tons of methane) in 2004, representing 9.0 percent of total U.S. greenhouse gas emissions. Methane emissions in 2004 were 0.9 percent (5.6 $MMTCO_2e$) higher than the 2003 level of 633.9 $MMTCO_2e$ (Table ES4). The increase is attributable primarily to increases in methane emissions from landfills and, to a lesser extent, emissions associated with animal waste, coal mines, and rice cultivation. Despite the 0.9percent increase in 2004, methane emissions still were 81.9 $MMTCO_2e$ (11.4 percent) below their 1990 level of 721.4 $MMTCO_2e$.

Methane emissions come from four source categories, three major and one minor. The major sources are energy, agriculture, and waste management, and the minor source is industrial processes. The three major sources accounted for 40, 29, and 31 percent, respectively, of total 2004 U.S. emissions of methane. Trends in the major sources of anthropogenic methane emissions since 1990 are illustrated in Figure ES4.

Energy Sources

Methane emissions from energy sources (coal mining, natural gas systems, petroleum systems, stationary combustion, and mobile source combustion) were nearly unchanged at 256.3 $MMTCO_2e$ in 2004 from the 2003 level of 256.4 $MMTCO_2e$. Methane emissions from energy sources have fallen by 6.8 percent since 1990; however, after bottoming out in 2001, methane emissions from energy sources rose in each of the next 2 years.

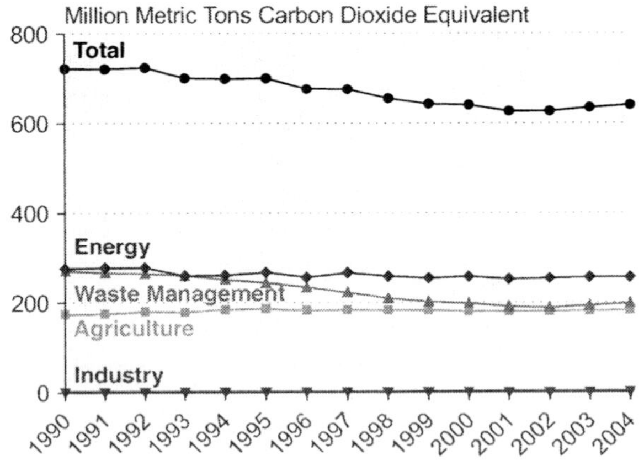

Source: Estimates presented in this report.

Figure ES4. U.S. Emissions of Methane by Source, 1990-2004

Agricultural Sources

Methane emissions from agricultural sources, at 182.3 $MMTCO_2e$, represented 29 percent of total U.S. anthropogenic methane emissions in 2004. Emissions increased by less than 1 percent (0.9 $MMTCO_2e$) in 2004 relative to 2003, as a small decrease in emissions from enteric fermentation in domesticated animals was offset by increases in emissions from animal waste management, rice cultivation, and burning of crop residues. Of the total 2004 methane emissions from agricultural sources, 93 percent (170.0 $MMTCO_2e$) resulted from livestock management, of which 68 percent (115.2 $MMTCO_2e$) was from enteric fermentation and the remainder (54.7 $MMTCO_2e$) was from anaerobic decomposition of livestock wastes. Methane emissions from rice cultivation (11 $MMTCO_2e$) and crop residue burning (1.3 $MMTCO_2e$) together represented about 6.7 percent of total methane emissions from agricultural sources in 2004, which have increased by 5.1 percent since 1990.

Waste Management

Methane emissions from waste management, which at 198.2 $MMTCO_2e$ accounted for 31 percent of U.S. anthropogenic methane emissions in 2004, were 2.4 percent above the 2003 level of 193.6 $MMTCO_2e$. Landfills—the largest single source of U.S. anthropogenic methane emissions—represented 92 percent (182.6 $MMTCO_2e$) of total U.S. methane emissions from waste management in 2004. The remainder (15.6 $MMTCO_2e$) of 2004 methane emissions from waste management was associated with domestic wastewater treatment. Methane emissions from waste management have fallen by 27 percent (72.0 $MMTCO_2e$) from their 1990 level of 270.2 $MMTCO_2e$, due largely to increased methane recovery at landfills, which increased to 122.1 $MMTCO_2e$ in 2004 from 21.7 $MMTCO_2e$ in 1990. Even at these higher methane recovery levels, however, waste management emissions, after bottoming out in 2002, increased in both 2003 and 2004.

Table ES4. U.S. Methane Emissions from Anthropogenic Sources, 1990 and 1996-2004 (Million Metric Tons Carbon Dioxide Equivalent)

Source	1990	1996	1997	1998	1999	2000	2001	2002	2003	P2004
Energy Sources										
Coal Mining	97.7	73.4	80.4	75.6	71.7	68.5	68.0	64.7	67.2	68.2
Natural Gas Systems	128.9	138.1	144.3	143.9	144.2	151.0	147.0	154.0	153.1	152.6
Petroleum Systems	29.9	26.4	26.3	25.5	24.0	23.8	23.7	23.5	23.3	23.2
Stationary Combustion	13.0	13.3	10.1	9.1	9.6	10.1	8.7	7.6	8.5	8.0
Mobile Sources	5.6	5.1	5.1	4.8	4.8	4.7	4.6	4.6	4.3	4.4
Total Energy Sources	**275.0**	**256.3**	**266.2**	**258.9**	**254.4**	**258.1**	**252.0**	**254.3**	**256.4**	**256.3**
Waste Management										
Landfills	257.0	219.8	208.1	195.0	187.4	183.3	175.8	173.4	178.1	182.6
Wastewater Treatment	13.2	14.3	14.5	14.7	14.8	15.0	15.2	15.3	15.5	15.6
Total Waste Management	**270.2**	**234.2**	**222.6**	**209.7**	**202.2**	**198.3**	**191.0**	**188.7**	**193.6**	**198.2**
Agricultural Sources										
Enteric Fermentation	119.6	122.0	119.1	117.2	117.3	116.3	115.1	115.7	116.2	115.2
Animal Waste	43.5	49.7	52.7	53.6	52.7	52.8	53.3	53.7	54.2	54.7
Rice Cultivation	9.3	9.4	10.3	10.7	11.5	10.2	10.7	10.2	9.8	11.0
Crop Residue Burning	1.0	1.1	1.1	1.1	1.1	1.1	1.1	1.0	1.2	1.3
Total Agricultural Sources	**173.4**	**182.2**	**183.3**	**182.6**	**182.5**	**180.5**	**180.3**	**180.6**	**181.3**	**182.3**
Industrial Processes	**2.7**	**3.1**	**3.1**	**3.1**	**3.1**	**2.9**	**2.5**	**2.6**	**2.6**	**2.7**
Total	**721.4**	**675.8**	**675.2**	**654.2**	**642.2**	**639.8**	**625.8**	**626.2**	**633.9**	**639.5**

P = preliminary data.

Notes: Data in this table are revised from the data contained in the previous EIA report, *Emissions of Greenhouse Gases in the United States 2003,* DOE/EIA0573(2003) (Washington, DC, December 2004). Totals may not equal sum of components due to independent rounding.

Sources: EIA estimates presented in this report. Emissions calculations based on Intergovernmental Panel on Climate Change, *Greenhouse Gas Inventory Reference Manual: Revised 1996 IPCC Guidelines for National Greenhouse Gas Inventories,* Vol. 3 (Paris, France, 1997), pp. 4.834.84, web site www.ipcc.ch/pub/guide.htm; and U.S. U.S. Environmental Protection Agency, *Inventory of U.S. Greenhouse Gas Emissions and Sinks 19902003,* EPA430R05003 (Washington, DC, April 2005), web site http://yosemite.epa.gov/oar/globalwarming.nsf/content/ResourceCenter PublicationsGHGEmissions USEmissionsInventory2005.html.

Industrial Processes

Methane emissions from industrial processes totaled 2.7 $MMTCO_2e$ in 2004, including 1.6 $MMTCO_2e$ from chemical manufacturing and 1.1 $MMTCO_2e$ from iron and steel production. Since 1990, growth in methane emissions from chemical production has been balanced by reductions in emissions from iron and steel production, leaving the 2004 total for industrial processes approximately equal to their 1990 level.

Nitrous Oxide

Estimated U.S. anthropogenic emissions of nitrous oxide in 2004 totaled 353.7 $MMTCO_2e$, or 1.2 MMT nitrous oxide. Nitrous oxide emissions represented 5.0 percent of total U.S. greenhouse gas emissions in 2004 and were 5.5 percent above the 2003 level of 335.2 $MMTCO_2e$. Most of the increase from 2003 can be attributed to increased emissions from agricultural sources, which rose by 17.4 $MMTCO_2e$ in 2004 and represented 94 percent of the overall increase in nitrous oxide emissions (18.5 $MMTCO_2e$) from 2003. The 2004 level marks the first year since 2000 in which U.S. emissions of nitrous oxide have been higher than the 1990 level of 337.0 $MMTCO_2e$ (Table ES5 and Figure ES5).

Agriculture

Agricultural sources, at 265.2 $MMTCO_2e$, accounted for 75 percent of total U.S. nitrous oxide emissions in 2004.

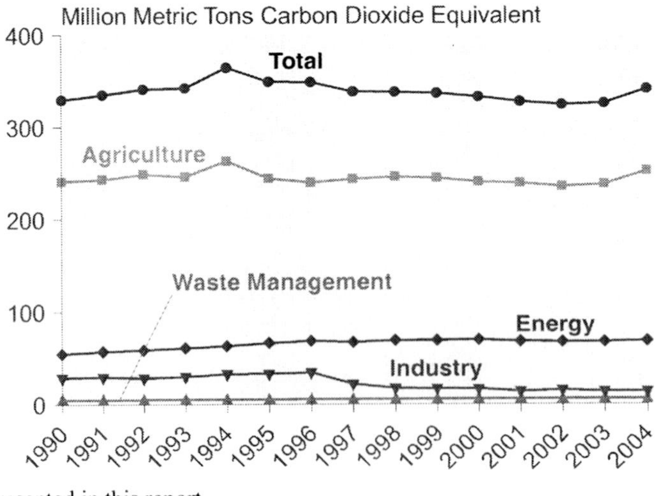

Source: Estimates presented in this report.

Figure ES5. U.S. Emissions of Nitrous Oxide by Source, 1990-2004

Agricultural emissions in 2004 were 7.0 percent above the 2003 total of 247.8 $MMTCO_2e$, primarily as the result of an increase of 9.5 percent (17.7 $MMTCO_2e$) in emissions from the nitrogen fertilization of agricultural soils. Emissions from nitrogen fertilization, at 204.3 $MMTCO_2e$, accounted for 77 percent of nitrous oxide emissions from agriculture in 2004. Emissions from the solid waste of domesticated animals, at 60.3 $MMTCO_2e$, made up 23 percent of agricultural nitrous oxide emissions in 2004, and burning of crop residues produced another 0.6 $MMTCO_2e$. Total U.S. emissions of nitrous oxide from agriculture sources have increased by 6.4 percent since 1990.

Table ES5. Estimated U.S. Emissions of Nitrous Oxide, 1990 and 1996-2004 (Million Metric Tons Carbon Dioxide Equivalent)

Source	1990	1996	1997	1998	1999	2000	2001	2002	2003	P2004
Agriculture										
Nitrogen Fertilization of Soils	186.9	184.2	190.5	194.0	192.7	189.4	189.1	185.1	186.6	204.3
Solid Waste of Domesticated Animals	61.9	65.1	63.8	62.9	62.4	61.8	61.4	61.2	60.7	60.3
Crop Residue Burning	0.5	0.5	0.5	0.6	0.5	0.6	0.6	0.5	0.5	0.6
Subtotal	**249.3**	**249.8**	**254.8**	**257.4**	**255.6**	**251.8**	**251.1**	**246.8**	**247.8**	**265.2**
Energy Use										
Mobile Combustion	41.1	53.9	52.4	54.2	54.3	54.3	53.3	53.0	53.0	53.8
Stationary Combustion	13.3	14.5	14.6	14.5	14.7	15.1	14.5	14.2	14.5	14.7
Subtotal	**54.4**	**68.5**	**67.0**	**68.7**	**68.9**	**69.4**	**67.9**	**67.2**	**67.5**	**68.4**
Industrial Sources	**28.6**	**34.3**	**21.8**	**17.2**	**16.8**	**16.6**	**14.0**	**15.2**	**14.0**	**14.0**
Waste Management										
Human Sewage in Wastewater	4.6	5.2	5.2	5.3	5.5	5.6	5.6	5.7	5.7	5.8
Waste Combustion	0.2	0.2	0.2	0.2	0.2	0.2	0.2	0.2	0.2	0.2
Subtotal	**4.8**	**5.4**	**5.4**	**5.5**	**5.7**	**5.8**	**5.9**	**5.9**	**6.0**	**6.0**
Total	**337.0**	**358.0**	**349.1**	**348.8**	**347.1**	**343.5**	**338.8**	**335.1**	**335.2**	**353.7**

P = preliminary data.

Notes: Data in this table are revised from the data contained in the previous EIA report, *Emissions of Greenhouse Gases in the United States 2003,* DOE/EIA0573(2003) (Washington, DC, December 2004). Totals may not equal sum of components due to independent rounding.

Sources: Estimates presented in this report. Emissions calculations based on Intergovernmental Panel on Climate Change, *Greenhouse Gas Inventory Reference Manual: Revised 1996 IPCC Guidelines for National Greenhouse Gas Inventories,* Vol. 3 (Paris, France, 1997), pp. 4.814.94, web site www.ipcc.ch/pub/guide.htm; and U.S. Environmental Protection Agency, *Inventory of U.S. Greenhouse Gas Emissions and Sinks: 19902003,* EPA 430R05003 (Washington, DC, April 2005), web site http://yosemite.epa.gov/oar/globalwarming.nsf/content/ResourceCenter PublicationsGHGEmissionsUSEmissionsInventory2005.html.

Energy Use

U.S. nitrous oxide emissions associated with fossil fuel combustion in 2004 were 68.4 $MMTCO_2e$, or 19 percent of total nitrous oxide emissions. Most of the energyrelated emissions of nitrous oxide in 2004 (79 percent or 53.8 $MMTCO_2e$) were from mobile sources, principally, motor vehicles equipped with catalytic converters. The remainder (21 percent or 14.7 $MMTCO_2e$) was from stationary source combustion of fossil fuels. Nitrous oxide emissions from energy sources have increased by 26 percent since 1990.

Industrial Processes and Waste Management

Industrial processes and waste management facilities were responsible for 3.9 percent and 1.6 percent, respectively, of total U.S. nitrous oxide emissions in 2004. Industrial process emissions in 2004 remained unchanged from 2003 at 14.0 $MMTCO_2e$. Industrial process emissions have fallen by 51 percent since 1990 due to decreases in nitrous oxide emissions from the manufacture of adipic acid.

Nitrous oxide emissions from waste management facilities in 2004 also were unchanged from their 2003 level, at 6.0 $MMTCO_2e$, including 5.8 $MMTCO_2e$ from human sewage in wastewater and 0.2 $MMTCO_2e$ from waste combustion. Nitrous oxide emissions from waste management facilities have increased by 25 percent since 1990.

Other Gases: Hyd rofluorocarbons, Perfluorocarbons, and Sulfur Hexafluoride

HFCs, PFCs, and SF6 are three classes of engineered gases that accounted for 2.2 percent of total U.S. greenhouse gas emissions in 2004. At 155.9 MMTCO_2e, their emissions were 9.6 percent higher than in 2003 (142.4 MMTCO_2e). The increase in emissions of the engineered gases from 2003 to 2004 resulted largely from a 12percent increase in HFC emissions, which more than offset decreases in emissions of PFCs (7.5 percent lower than in 2003) and SF6 (2.2 percent lower) (Table ES6).

Table ES6. U.S. Emissions of Hydrofluorocarbons, Perfluorocarbons, and Sulfur Hexafluoride, 1990 and 1996-2004 (Million Metric Tons Carbon Dioxide Equivalent)

Gas	1990	1996	1997	1998	1999	2000	2001	2002	2003	P2004
Hydrofluorocarbons										
HFC-23	36.1	32.3	31.2	41.7	31.7	30.9	20.6	20.6	13.0	16.3
HFC-125	0.0	6.3	8.6	10.7	12.1	13.6	14.9	16.2	17.8	19.7
HFC-134a	0.0	24.5	31.4	36.7	42.2	48.0	52.7	56.9	60.5	65.8
HFC-143a	0.0	2.3	4.0	5.9	7.5	9.3	11.4	13.8	16.5	19.5
HFC-236fa	0.0	0.0	0.1	0.6	1.3	2.0	2.6	3.2	3.5	3.5
Total HFCs	**36.1**	**65.5**	**75.1**	**95.5**	**94.9**	**103.8**	**102.2**	**110.7**	**111.3**	**124.8**
Perfluorocarbons										
CF_4	14.8	11.0	10.1	8.7	8.6	8.7	4.2	5.0	3.8	3.2
C_2F_6	4.8	5.4	5.6	5.9	6.0	5.0	3.4	3.8	3.4	3.5
Total PFCs	**19.6**	**16.4**	**15.7**	**14.6**	**14.6**	**13.7**	**7.6**	**8.8**	**7.2**	**6.7**
Other HFCs, PFCs/PFPEs	**0.3**	**3.4**	**4.5**	**5.5**	**6.2**	**6.5**	**6.7**	**7.1**	**7.6**	**8.6**
Sulfur Hexafluoride	**32.1**	**29.1**	**26.7**	**22.1**	**21.7**	**18.1**	**17.4**	**16.4**	**16.3**	**16.0**
Total Emissions	**88.1**	**114.3**	**122.0**	**137.7**	**137.4**	**142.1**	**133.9**	**143.1**	**142.4**	**155.9**

P = preliminary data.

Notes: Other HFCs, PFCs/PFPEs include HFC152a, HFC227ea, HFC245fa, HFC4310mee, and a variety of PFCs and perfluoropolyethers (PFPEs). They are grouped together to protect confidential data. Totals may not equal sum of components due to independent rounding.

Source: U.S. Environmental Protection Agency, Office of Air and Radiation, web site www.epa.gov/globalwarming/ (preliminary estimates, September 2005).

At 124.8 MMTCO_2e, emissions of HFCs made up the majority of U.S. emissions of engineered greenhouse gases, followed by SF6 at 16.0 MMTCO_2e and PFCs at 6.7 MMTCO_2e. Another group of engineered gases, consisting of other HFCs, other PFCs, and perfluoropolyethers (PFPEs), includes HFC152a, HFC227ea, HFC245fa, HFC4310mee, and a variety of PFCs and PFPEs. They are grouped together in this report to protect confidential data. In 2004, their combined emissions totaled 8.6 MMTCO_2e. Emissions of the engineered gases in this group in 2004 were 13 percent higher than in 2003 and an order of magnitude higher than in 1990, when they totaled less than 0.3 MMTCO_2e. Since 1990, HFC emissions from U.S. sources have increased by 246 percent, PFC emissions have fallen by 66 percent, and SF6 emissions have fallen by 50 percent.

Emissions of the highGWP gases specified in the Kyoto Protocol are very small (at most a few thousand metric tons). On the other hand, some of the gases (including PFCs and SF_6) have atmospheric lifetimes measured in the thousands of years, and consequently they are potent greenhouse gases with GWPs thousands of times higher than that of carbon dioxide per unit of molecular weight. Some of the commercially produced HFCs (134a, 152a, 4310mee, and 227ea), which are used as replacements for chlorofluorocarbons (CFCs) and hydrochlorofluorocarbons (HCFCs), have shorter atmospheric lifetimes, ranging from 1 to 33 years.

Land-use Change and Forestry

Forest lands in the United States are net absorbers of carbon dioxide from the atmosphere, primarily as a result of the reversal of the extensive deforestation that occurred in the United States during the late 19th and early 20th centuries. Since then, millions of acres of formerly cultivated land have been abandoned and have returned to forest, with the regrowth of forests sequestering carbon on a large scale. The process is steadily diminishing, however, because the rate at which forests absorb carbon slows as the trees mature, and because the rate of reforestation has slowed.

The U.S. Environmental Protection Agency (EPA) estimates annual U.S. carbon sequestration from land-use change and forestry in 2003 at 828.0 $MMTCO_2e$,[9] representing an offset of 11.9 percent of total 2003 U.S. greenhouse gas emissions (6,983.2 $MMTCO_2e$). In 1990, carbon sequestration attributable to land use and forestry was 1,042.1 $MMTCO_2e$, or 16.9 percent of total 1990 U.S. greenhouse gas emissions (6,148.8 $MMTCO_2e$).[10] The EPA's 2003 estimates for carbon sequestration from land-use change and forestry include 752.7 $MMTCO_2e$ from forested land, 58.7 $MMTCO_2e$ from urban trees, 10.1 $MMTCO_2e$ from landfilled yard trimmings and food scraps, and 6.6 $MMTCO_2e$ from agricultural soils (Table ES7).

Uncertainty in Emissions Estimates

The emissions numbers presented in this report are estimates based on estimated activity data and estimated emission factors. As such, they have an element of uncertainty, given that the activity data and emission factors on which the emission estimates are based also have a range of possible values. The activity data and emission factors can themselves be characterized by systematic biases and/or random errors.

Table ES7. Net Carbon Dioxide Sequestration from U.S. Land-use Change and Forestry, 1990 and 19972003 (Million Metric Tons Carbon Dioxide Equivalent)

Component	1990	1997	1998	1999	2000	2001	2002	2003
Forest Land Remaining Forest Land: Changes in Forest Carbon Stocks	949.3	851.0	805.5	751.7	747.9	750.9	751.5	752.7
Cropland Remaining Cropland: Changes in Agricultural Soil Carbon Stocks	8.1	7.4	4.3	4.3	5.7	7.1	6.2	6.6
Settlements Remaining Settlements	84.7	71.6	71.2	70.0	68.9	68.9	68.8	68.7
Urban Trees	*58.7*	*58.7*	*58.7*	*58.7*	*58.7*	*58.7*	*58.7*	*58.7*
Landfilled Yard Trimmings and Food Scraps	*26.0*	*12.9*	*12.5*	*11.4*	*10.2*	*10.3*	*10.2*	*10.1*
Total Net Flux	**1,042.1**	**930.0**	**881.0**	**826.0**	**822.5**	**826.9**	**826.5**	**828.0**

Note: Totals may not equal sum of components due to independent rounding.

Source: U.S. Environmental Protection Agency, *Inventory of U.S. Greenhouse Gas Emissions and Sinks: 19902003,* EPA 430R05003 (Washington, DC, April 2005), web site http://yosemite.epa.gov/oar/globalwarming.nsf/content/ResourceCenter PublicationsGHGEmissionsUSEmissionsInventory2005.html.

In 2000, EIA employed a Monte Carlo analysis to estimate the range of uncertainty, at a 95percent confidence level, around estimated emissions of carbon dioxide, methane, and nitrous oxide (HFCs, PFCs, and SF6 were not part of the analysis).[11]

The Monte Carlo simulations revealed that uncertainty varies by type of gas. There is less uncertainty around the simulated mean for carbon dioxide (1.4 percent to +1.3 percent) than for methane (15.6 percent to 16.0 percent) or nitrous oxide (53.5 percent to +54.2 percent). The simulations also showed that the uncertainty around the simulated mean of total greenhouse gas emissions (excluding HFCs, PFCs, and SF_6) is 4.4 percent to +4.6 percent.

The reliability of emissions data varies by category and by source. In general, the estimates for carbon dioxide emissions are more reliable than the estimates for other gases. It is likely that the estimate of carbon dioxide emissions is accurate to within 5 percent. Estimates of methane emissions are much more uncertain, with a level of uncertainty that may exceed 30 percent. Estimates of methane emissions are also likely to understate actual emissions as a result of the exclusion of sources that are unknown or difficult to quantify, such as abandoned coal mines and industrial wastewater. Nitrous oxide emissions estimates are much less reliable than those for carbon dioxide or methane emissions, in part because nitrous oxide emissions have been studied far less than emissions of the other greenhouse gases and in part because the largest apparent sources of nitrous oxide emissions are area sources that result from biological activity, which makes for emissions that are highly variable and hard to measure or characterize. The uncertainty for nitrous oxide emissions may exceed 100 percent.

About This Report

The Energy Policy Act of 1992 requires the Energy Information Administration (EIA) to prepare an inventory of aggregate U.S. national emissions of greenhouse gases for the period 19871990, with annual updates thereafter. This report contains data from the twelfth annual inventory update, covering national emissions over the period 19902003, with preliminary estimates of emissions for 2004.

EIA continually reviews its methods for estimating emissions of greenhouse gases. As better methods and information become available, EIA revises both current and historical emissions estimates (see "What's New," below).

This introductory chapter provides background information on U.S. greenhouse gases in a global context, the greenhouse effect and global climate change, and recent domestic and international developments to address climate change. Chapters 2 through 4 cover emissions of carbon dioxide, methane, and nitrous oxide, respectively. Chapter 5 focuses on emissions of engineered gases, including hydrofluorocarbons, perfluorocarbons, and sulfur hexafluoride. Chapter 6 describes potential sequestration and emissions of greenhouse gases as a result of land-use changes.

What's New

Carbon Dioxide

In this year's report, data on carbon dioxide emissions have been enhanced to include emissions from nonfuel uses of petroleum products, natural gas, and coal (see Table 12 in

Chapter 2). Emissions from nonfuel uses of energy fuels were included as part of energyrelated carbon dioxide emissions in previous editions of this report but were not shown separately. Emissions from nonfuel uses are now shown separately under energy consumption.

Methane

The method for estimating methane emissions from landfills uses data on total waste generated and waste landfilled, published in *Biocycle* magazine. Before this year, the data from *Biocycle* were available with a oneyear time lag, and EIA arrived at an estimate for the current year by scaling the most recent *Biocycle* data year according to changes in annual gross domestic product. For this year's report, *Biocycle* data were unavailable, creating a twoyear time lag. Thus, to ensure a more rigorous statistical approach, EIA revised 2003 and 2004 waste generation data by using a regression equation that correlated changes in waste generation since 1988 to changes in gross domestic product over the same period. This change had a negligible impact on the emissions estimates.

In addition, EIA has revised its estimates of methane recovered from landfills and used for energy production. Previous editions of this report erroneously included the avoided emissions of carbon dioxide from fossil fuel combustion displaced by landfill gastoenergy production in the estimate of methane recovered. This resulted in double counting of the impacts of fossil fuel displacement by landfillgastoenergy. For this year's report, avoided emissions have been removed from the estimates of methane emissions from landfills going back to 1990. EIA has also revised estimates of methane emissions from swine waste by adding detail on animal size.

Other Gases: HFCs, PFCs, and SF6

The U.S. Environmental Protection Agency (EPA), Office of Air and Radiation, has made revisions to the data and estimation methodologies used for other gases—hydrofluorocarbons (HFCs), perfluorocarbons (PFCs), and sulfur hexafluoride (SF_6)—in its most recent emissions inventory.[1] Those changes are reflected in the EPA's historical emissions estimates, as described below:

- **Electricity Transmission and Distribution.** The changes in calculations of emissions from electricity transmission and distribution are the result of incorporating more uptodate transmission mileage data and the inclusion of additional historical partner data in the EPA's SF6 Emissions Reduction Partnership for Electric Power Systems for 2000, 2001, and 2002. Previously, the 2001 Utility Data Institute (UDI) Directory of Electric Power Producers and Distributors was used by the EPA to estimate SF6 emissions for 2001 and 2002. Those numbers have been revised to account for increases in transmission mileage during 2001 and 2002, primarily as a result of growth in the U.S. transmission system. Accordingly, estimates of nonpartner and nonreporting partner emissions have been recalculated in the nonreporting partner regression equations. Because transmission miles are highly correlated with SF6 emissions, the EPA has used these regression equations to calculate emissions from nonpartners and nonreporting partners in the SF6 emissions reduction partnership. In addition to transmission mileage revisions, the electric power system emission estimates have also been recalculated, based on additional historical partner data. Specifically, the regression equations for each respective year of the historical partner submissions have been updated, resulting in new extrapolations to nonreporting partners as well. These revisions resulted in an average annual decrease in estimated SF6 emissions from electric power systems of 0.2 percent, or less than 0.1 million metric tons carbon dioxide equivalent ($MMTCO_2e$) for the 20002002 period.

- **Magnesium Production and Processing.** The emissions estimates in this report have been revised to reflect new historical data supplied by the U.S. Geological Survey and participants in the EPA's SF6 Emission Reduction Partnership for the Magnesium Industry. This change resulted in an average annual increase in estimated SF6 emissions from magnesium production and processing of less than 0.1 $MMTCO_2e$ (4.1 percent) for the 20002002 period.
- **Substitution of OzoneDepleting Substances.** The EPA has updated assumptions for its Vintaging Model pertaining to market trends in chemicals and chemical substitutes. These changes resulted in an average annual net increase in estimated HFC and PFC emissions of less than 0.1 $MMTCO_2e$ (4.1 percent) for the 19902002 period.
- **Aluminum Production.** As the result of an EPAfunded study, facilityspecific slope coefficients for three U.S. aluminum smelters have been reestimated. The new coefficients have been used by the EPA in place of the Intergovernmental Panel on Climate Change (IPCC) defaults for revising the appropriate smelterspecific emission factors. The EPA provided the revised data to EIA, along with additional recently reported data concerning smelter operating parameters by participants in the EPA's Voluntary Aluminum Industrial Partnership Program. These changes resulted in an average annual increase of less than 0.1 $MMTCO_2e$ (0.2 percent) for the 19902002 period.
- **HCFC22 Production.** The EPA has used reports from the Alliance for Responsible Atmospheric Policy to adjust the historical time series for HFC emissions from HCFC22 production. These changes resulted in an average annual decrease in HFC emissions from HCFC22 production of less than 0.1 $MMTCO_2e$ (0.01 percent) for the 19902002 period. The newer U.S. Department of Agriculture (USDA) Forestry Inventory and Analysis Database (FIADB) datasets[4] were considered for nonsoil forest carbon estimates, along with USDA Resources Planning Act (RPA) data.[5] This year, final values for all carbon pools were extrapolated from the last carbon stock change values calculated from the Forest Inventory and Analysis (FIA) survey data.

Land-use Issues

The carbon sequestration estimates in this year's report reflect several categorical and methodological changes from previous years that have been implemented by the EPA. They include improvements in calculations as well as changes made in accordance with new guidelines from the IPCC, published in its *Good Practice Guidance for Land Use, Land-use Change and Forestry* (LULUCF GPG).[2] Key changes in this year's inventory include the following:

- As recommended in the LULUCF GPG, carbon stocks are reported according to several land-use types and conversions—for example, forest land remaining forest land, nonforest land becoming forest land, and forest land becoming nonforest land.
- Changes have been made in the definitions of different forest carbon pools. Standing dead trees are now part of the "dead wood" pool. Above and belowground portions of various pools are now aggregated as above and belowground biomass pools. The soil pool and the forest floor are now classified as "soil organic carbon" and "litter," respectively.
- The estimate of organic carbon in soils in the conterminous United States was calculated from the State Soil Geographic database,[3] and data gaps were filled with representative values for similar soils.
- The newer U.S. Department of Agriculture (USDA) Forestry Inventory and Analysis Database (FIADB) datasets[4] were considered for nonsoil forest carbon estimates, along with USDA Resources Planning Act (RPA) data.[5]
- This year, final values for all carbon pools were extrapolated from the last carbon stock change values calculated from the Forest Inventory and Analysis (FIA) survey data.

U.S. Emissions in a Global Perspective

This report estimates that U.S. energyrelated carbon dioxide emissions in 2002 totaled 5,746 million metric tons (MMT). To put U.S. emissions in a global perspective, total energyrelated carbon dioxide emissions for the world in 2002 are estimated at 24,405 MMT, making U.S. emissions about 24 percent of the world total (Table 1).[6] Emissions for the mature market economies (North America, Western Europe, Japan, and Australia/New Zealand) in 2002 are estimated at 11,872 MMT, or about 49 percent of the world total. In 2002, the remaining 51 percent of worldwide energyrelated carbon dioxide emissions came from emerging economies (9,408 MMT) and the transitional economies of the former Soviet Union and Eastern Europe (3,124 MMT). By 2025, however, the U.S. share of total world emissions is projected to fall to 20 percent (7,587 MMT out of a global total of 38,396 MMT). The reason for the expected decline in the U.S. share is that U.S. energyrelated carbon dioxide emissions are projected to increase at an annual rate of 1.2 percent, while emissions from the emerging and transitional economies are projected to grow at annual rates of 3.2 and 1.5 percent, respectively.

The Greenhouse Effect and Global Climate Change

The Earth is warmed by radiant energy from the Sun. Over time, the amount of energy transmitted to the Earth's surface is equal to the amount of energy reradiated back into space in the form of infrared radiation, and the temperature of the Earth's surface stays roughly constant; however, the temperature of the Earth is strongly influenced by the existence, density, and composition of its atmosphere. Many gases in the Earth's atmosphere absorb infrared radiation reradiated from the surface, trapping heat in the lower atmosphere. Without the natural greenhouse effect, it is likely that the average temperature of the Earth's surface would be on the order of 19° Celsius, rather than the +14° Celsius actually observed.[7] The gases that help trap the Sun's heat close to the Earth's surface are referred to as "greenhouse gases." All greenhouse gases absorb infrared radiation (heat) at particular wavelengths.

The most important greenhouse gases are water vapor (H_2O), carbon dioxide (CO_2), methane (CH_4), nitrous oxide (N_2O), and several engineered gases, such as HFCs, PFCs, and SF_6. Water vapor is by far the most common, with an atmospheric concentration of nearly 1 percent, compared with less than 0.04 percent for carbon dioxide. The effect of human activity on global water vapor concentrations is considered negligible, however, and anthropogenic (humanmade) emissions of water vapor are not factored into national greenhouse gas emission inventories for the purposes of meeting the requirements of the United Nations Framework Convention on Climate Change (UNFCCC) or the Kyoto Protocol.[8] Concentrations of other greenhouse gases, such as methane and nitrous oxide, are a fraction of that for carbon dioxide (Table 2).

Scientists recognized in the early 1960s that concentrations of carbon dioxide in the Earth's atmosphere were increasing every year. Subsequently, they discovered that atmospheric concentrations of methane, nitrous oxide, and many engineered greenhouse gas chemicals also were rising. Because current concentrations of greenhouse gases keep the Earth at its present temperature, scientists began to postulate that increasing concentrations of greenhouse gases would make the Earth warmer. In computerbased simulation models, rising concentrations of greenhouse gases nearly always produce an increase in the average temperature of the Earth. Rising temperatures may, in turn, produce changes in weather and in the level of the oceans that might prove disruptive to current patterns of land use and human settlement, as well as to existing ecosystems. To date, however, it has proven difficult to disentangle the human impact on climate from normal temporal and spatial variations in temperature on both a global scale and geologic timeframe. The most recent report of the IPCC, an international assemblage of scientists commissioned by the United Nations to assess the scientific, technical, and socioeconomic information relevant for the understanding of the risk of humaninduced climate change, estimates that the global average surface temperature has increased by 0.6 ± 0.2°C since the late 19th century.[9] The IPCC goes on to conclude that: "There is new and stronger evidence that most of the warming observed over the last 50 years is attributable to human activities."[10]

In the aftermath of the IPCC report, the Domestic Policy Council, in May 2001 as part of its review of U.S. policy on climate change, requested that the National Academy of Sciences identify areas of uncertainty in the science of climate change, as well as review the IPCC report and summaries.[11] The National Academy of Sciences commissioned the National Research Council to carry out this review. The National Research Council in issuing its findings appeared to agree with some of the IPCC conclusions, but also seemed to suggest that further work needs to be done in identifying the impacts of natural climatic variability and reducing the uncertainty inherent in climate change modeling. Among the National Research Council findings are the following:[12]

> *Greenhouse gases are accumulating in Earth's atmosphere as a result of human activities, causing surface air temperatures and subsurface ocean temperatures to rise. Temperatures are, in fact, rising. The changes observed over the last several decades are likely mostly due to human activities, but we cannot rule out that some significant part of these changes is also a reflection of natural variability.*

Because there is considerable uncertainty in current understanding of how the climate system varies naturally and reacts to emissions of greenhouse gases and aerosols, current estimates of the magnitude of future warming should be regarded as tentative and subject to future adjustments (either upward or downward).

The committee generally agrees with the assessment of humancaused climate change presented in the IPCC Working Group I (WGI) scientific report, but seeks here to articulate more clearly the level of confidence that can be ascribed to those assessments and the caveats that need to be attached to them.

Table 1. World Carbon Dioxide Emissions by Region, 1990-2025 (Million Metric Tons Carbon Dioxide)

Region/Country	History			Projections				Average Annual Percent Change, 2002-2025
	1990	2001	2002	2010	2015	2020	2025	
Mature Market Economies								
North America	**5,778**	**6,660**	**6,696**	**7,478**	**7,934**	**8,417**	**8,984**	**1.3**
United States[a]	4,997	5,728	5,746	6,365	6,718	7,119	7,587	1.2
Canada	473	573	588	681	726	757	807	1.4
Mexico	308	359	363	432	490	541	591	2.1
Western Europe	**3,413**	**3,585**	**3,549**	**3,674**	**3,761**	**3,812**	**3,952**	**0.5**
Mature Market Asia	**1,284**	**1,610**	**1,627**	**1,731**	**1,780**	**1,822**	**1,852**	**0.6**
Japan	990	1,182	1,179	1,211	1,232	1,240	1,242	0.2
Australia/New Zealand	294	429	448	520	548	582	610	1.4
Total Mature Market	**10,474**	**11,855**	**11,872**	**12,883**	**13,475**	**14,051**	**14,788**	**1.0**
Transitional Economies								
Former Soviet Union	**3,798**	**2,393**	**2,399**	**2,804**	**3,040**	**3,201**	**3,379**	**1.5**
Russia	1,837	1,431	1,409	1,615	1,715	1,799	1,864	1.2
Other FSU	1,962	962	990	1,189	1,325	1,401	1,515	1.9
Eastern Europe	**1,095**	**744**	**726**	**839**	**898**	**951**	**1,006**	**1.4**
Total Transitional	**4,894**	**3,137**	**3,124**	**3,643**	**3,937**	**4,151**	**4,386**	**1.5**
Emerging Economies								
Emerging Asia	**3,890**	**5,967**	**6,205**	**9,306**	**10,863**	**12,263**	**13,540**	**3.5**
China	2,262	3,176	3,322	5,536	6,506	7,373	8,133	4.0
India	583	1,009	1,025	1,369	1,581	1,786	1,994	2.9
South Korea	234	431	451	549	623	676	723	2.1
Other Asia	811	1,351	1,407	1,853	2,154	2,428	2,689	2.9
Middle East	**845**	**1,311**	**1,361**	**1,761**	**1,975**	**2,163**	**2,352**	**2.4**
Africa	**655**	**840**	**854**	**1,122**	**1,283**	**1,415**	**1,524**	**2.5**
Central and South America	**711**	**998**	**988**	**1,289**	**1,480**	**1,639**	**1,806**	**2.7**
Brazil	250	343	342	433	502	583	679	3.0
Other Central/South America	461	655	646	856	979	1,056	1,128	2.5
Total Emerging	**6,101**	**9,116**	**9,408**	**13,478**	**15,602**	**17,480**	**19,222**	**3.2**
Total World	**21,469**	**24,108**	**24,405**	**30,005**	**33,014**	**35,682**	**38,396**	**2.0**

[a]Includes the 50 States and the District of Columbia.

Note: The U.S. numbers include carbon dioxide emissions attributable to renewable energy sources.

Sources: History: Energy Information Administration (EIA), *International Energy Annual 2002,* DOE/EIA02 19(2002) (Washington, DC, March 2004), web site www.eia.doe.gov/iea/; and data presented in this report. Projections: EIA, *Annual Energy Outlook 2006 Early Release,* DOE/EIA0383(2006) (Washington, DC, December 12, 2005), Table 18, web site www.eia.doe.gov/oiaf/aeo/ pdf/aeotab_1 8.pdf; and *International Energy Outlook 2005,* DOE/EIA0484(2005) (Washington, DC, February 2005), Table A10.

Additionally, the U.S. Academy, along with the science academies of 10 other countries, in June 2005 released a joint statement on climate change (see text box on page 6).

Greenhouse Gas Sources and Sinks

Most greenhouse gases have both natural and humanmade emission sources, and there are significant natural mechanisms (landbased or oceanbased "sinks") for removing them from the atmosphere; however, increased levels of anthropogenic emissions have pushed the total level of greenhouse gas emissions (both natural and anthropogenic) above their natural absorption rates. The positive imbalance between emissions and absorption has resulted in the continuing growth in atmospheric concentrations of these gases. Table 3 illustrates the relationship between anthropogenic and natural emissions and absorption of the principal greenhouse gases on an annual average basis during the 1990s.

Table 2. Global Atmospheric Concentrations of Selected Greenhouse Gases

	Carbon Dioxide	Methane	Nitrous Oxide	Sulfur Hexafluoride	Perfluoro-methane
Item	(parts per million)			(parts per trillion)	
Pre-industrial (1750) Atmospheric Concentration. .	278	0.700	0.270	0	40
1998 Atmospheric Concentration.	365	1.745	0.314	4.2	80
Average Annual Change, 1990-1999.	1.5[a]	0.007[a]	0.0008	0.2	1.0
Atmospheric Lifetime (Years).	50–200[b]	12[c]	114[c]	3,200	>50,000

[a]The rate has fluctuated between 0.9 and 2.8 parts per million per year for CO2 and between 0 and 0.013 parts per million per year for methane over the 19901 999 period.

bNo single lifetime can be defined for CO2 because uptake rates differ for different removal processes.

[c]This lifetime has been defined as an "adjustment time" that takes into account the indirect effect of the gas on its own residence time.

Source: Intergovernmental Panel on Climate Change, *Climate Change 2001: The Scientific Basis* (Cambridge, UK: Cambridge University Press, 2001), pp. 38 and 244.

Water Vapor. Water vapor, as noted above, is the most common greenhouse gas present in the atmosphere. It is emitted into the atmosphere in enormous volumes through natural evaporation from oceans, lakes, and soils and is returned to Earth in the form of rain and snow. The recent IPCC report, however, cites a possible positive feedback from increased water vapor formation due to increased warming caused by rising atmospheric concentrations of carbon dioxide.[13] Elevated atmospheric temperatures increase the waterholding capability of the atmosphere. According to some of the IPCC emission scenarios, higher water vapor content could double the predicted atmospheric warming from what it would be if the water vapor concentration stayed constant. These scenarios, however, have an element of uncertainty due to the possible countervailing effect of increased cloud formation, which can act to cool the planet by absorbing and reflecting solar radiation or warm the planet through the emission of longwave radiation. According to the IPCC, increases in atmospheric temperatures would not necessarily result in increased concentrations of water vapor, because most of the atmosphere today is undersaturated.

Summary of Science Academies' Joint Statement on Global Response to Climate Change

The U.S. National Academy of Sciences and the national science academies of Brazil, Canada, China, France, Germany, India, Italy, Japan, Russia, and the United Kingdom in June

2005 issued a joint statement on the global response to climate change. Their statement stressed the following principles:

- Climate change is real: warming attributable to human activities is taking place and has already led to changes in the Earth's climate.
- The causes of climate change should be reduced: nations should identify costeffective steps that they can take now to reduce the growth of greenhouse gas emissions.
- Nations should prepare for the consequences of climate change: because large portions of the climate respond slowly, even if steps were taken today to stabilize emissions at current levels the climate would continue to change; therefore, nations should prepare for unavoidable changes.
- An international study should be launched: the G8 nations should develop a sciencebased approach to targets for greenhouse gas emissions that avoid "unacceptable" impacts.
- There should be cooperation with developing countries: G8 nations should work with developing countries to find solutions that best fit the circumstances of those countries.

Table 3. Global Natural and Anthropogenic Sources **and Absorption of Greenhouse Gases in the 1990s**

Gas	Sources			Absorption	Annual Increase in Gas in the Atmosphere
	Natural	Human-Made	Total		
Carbon Dioxide (Million Metric Tons of Gas)[a]	770,000	23,100	793,100	781,400	11,700
Methane (Million Metric Tons of Gas)[b]	239	359	598	576	22
Nitrous Oxide (Million Metric Tons of Gas)[c]	9.5	6.9	16.4	12.6	3.8

[a]Carbon dioxide natural source and absorption of 770,000 million metric tons carbon dioxide, based on balanced flux of 40,000 million metric tons between land and atmosphere and 330,000 million metric tons between oceans and atmosphere, from Intergovernmental Panel on Climate Change, *Climate Change 2001: The Scientific Basis* (Cambridge, UK: Cambridge University Press, 2001), Figure 3.1, p. 188. Humanmade emissions of 23,100 million metric tons and distribution of those emissions (atmospheric absorption 11,700 million metric tons, ocean absorption 6,200 million metric tons, and land absorption 5,100 million metric tons), taken from Intergovernmental Panel on Climate Change, *Climate Change 2001: The Scientific Basis* (Cambridge, UK: Cambridge University Press, 2001), p. 39.

[b]Methane total sources, absorption, and annual atmospheric increases from Intergovernmental Panel on Climate Change, *Climate Change 2001: The Scientific Basis* (Cambridge, UK: Cambridge University Press, 2001), Table 4.2, p. 250. Distinction between natural and humanmade sources based on the assumption that 60 percent of total sources are anthropogenic, from Intergovernmental Panel on Climate Change, *Climate Change 2001: The Scientific Basis* (Cambridge, UK: Cambridge University Press, 2001), p. 248.

[c]Nitrous oxide total and humanmade sources, absorption, and atmospheric increases from Intergovernmental Panel on Climate Change, *Climate Change 2001: The Scientific Basis* (Cambridge, UK: Cambridge University Press, 2001), Table 4.4, p. 252. Nitrous oxide natural sources (9.5 million metric tons of gas) derived by subtracting humanmade sources from total sources.

Source: Intergovernmental Panel on Climate Change, *Climate Change 2001: The Scientific Basis* (Cambridge, UK: Cambridge University Press, 2001).

Carbon Dioxide. Carbon is a common element on the planet, and immense quantities can be found in the atmosphere, in soils, in carbonate rocks, and dissolved in ocean water. All life on Earth participates in the "carbon cycle," by which carbon dioxide is extracted from the air by plants and decomposed into carbon and oxygen, with the carbon being incorporated into plant biomass and the oxygen released to the atmosphere. Plant biomass, in turn, ultimately decays (oxidizes), releasing carbon dioxide back into the atmosphere or storing organic carbon in soil or rock. There are vast exchanges of carbon dioxide between the ocean and the atmosphere, with the ocean absorbing carbon from the atmosphere and plant life in the ocean absorbing carbon from water, dying, and spreading organic carbon on the sea bottom, where it is eventually incorporated into carbonate rocks such as limestone.

Records from Antarctic ice cores indicate that the carbon cycle has been in a state of imbalance for the past 200 years, with emissions of carbon dioxide to the atmosphere exceeding absorption. Consequently, carbon dioxide concentrations in the atmosphere have been rising steadily. Because of its relative abundance, total carbon dioxide in the atmosphere has a radiative forcing value of 1.46 watts per square meter.[14] (See page 10 for a discussion of radiative forcing.) According to the IPCC, before 1750, atmospheric carbon dioxide concentration was around 280 ± 10 parts per million for several thousand years. The IPCC goes on to say that the present carbon dioxide concentration has not been exceeded during the past 420,000 years, and likely not during the past 20 million years.[15] The most important natural sources of carbon dioxide are releases from the oceans (330 billion metric tons of carbon dioxide per year) and land (440 billion metric tons of carbon dioxide annually), including 220 billion metric tons of carbon dioxide from plant respiration, 202 billion metric tons of carbon dioxide from nonplant respiration (bacteria, fungi, and herbivores), and 15 billion metric tons of carbon dioxide from combustion in natural and humanmade fires.[16] Known anthropogenic sources (including deforestation) were estimated to account for about 29 billion metric tons of carbon dioxide per year during the 1989 to 1998 time period.[17] The principal anthropogenic source is the combustion of fossil fuels, which accounts for about 80 percent of total anthropogenic emissions of carbon worldwide. Natural processes—primarily, uptake by the ocean and photosynthesis—absorb substantially all the naturally produced carbon dioxide and some of the anthropogenic carbon dioxide, leading to an annual net increase in carbon dioxide in the atmosphere of 11.4 to 12.1 billion metric tons.[18]

Methane. Methane is also a common compound. The methane cycle is less well understood than the carbon cycle. Natural methane is released primarily by anaerobic decay of vegetation in wetlands, by the digestive tracts of termites in the tropics, by the ocean, and by leakage from methane hydrate deposits. The principal anthropogenic sources are leakages from the production of fossil fuels, humanpromoted anaerobic decay in landfills, and the digestive processes of domestic animals. Anthropogenic sources are estimated to account for 60 percent of total methane emissions.[19] The main sources of absorption are thought to be tropospheric reactions with hydroxyl (OH) radicals that break down methane into the methyl radical CH3 and water vapor (506 MMT methane), stratospheric reactions with hydroxyl radicals and chlorine (40 MMT methane), and decomposition by bacteria in soils (30 MMT methane). Known and unknown sources of methane are estimated to total 598 MMT annually; known sinks (i.e., absorption by natural processes) total about 576 MMT. The annual increase in methane concentration in the atmosphere accounts for the difference of 22 MMT.[20] The radiative forcing of methane is 0.48 watts per square meter, about onethird that of carbon dioxide.[21]

Nitrous Oxide. The sources and absorption of nitrous oxide are much more speculative than those for other greenhouse gases. The principal natural sources are thought to be bacterial breakdown of nitrogen compounds in soils, particularly forest soils, fluxes from ocean upwellings, and stratospheric photo dissociation and reaction with electronically excited oxygen atoms. The primary humanmade sources are enhancement of natural processes through application of nitrogen fertilizers, combustion of fuels (in fossilfueled power plants and from the catalytic converters in automobiles), certain industrial processes (nylon and nitric acid production), biomass burning, and cattle and feedlots. Worldwide, estimated known sources of nitrous oxide total 16.4 MMT annually (6.9 MMT from anthropogenic sources), and known sinks total 12.6 MMT. The annual increase in concentrations in the atmosphere is thought to total 3.8 MMT.[22] The radiative forcing of nitrous oxide is 0.15 watts per square meter, about onetenth that of carbon dioxide.[23]

Halocarbons and Other Gases. During the 20th century, human ingenuity created an array of "engineered" chemicals, not normally found in nature, whose special characteristics render them particularly useful. One family of engineered gases is the halocarbons. A halocarbon is a compound containing carbon and either chlorine, bromine, or fluorine. Halocarbons are powerful greenhouse gases. Halocarbons that contain bromine or chlorine also deplete the Earth's ozone layer.

One of the bestknown groups of halocarbons is the chlorofluorocarbons (CFCs), particularly CFC12, often known by its trade name, "Freon12." CFCs have many desirable features: they are relatively simple to manufacture, inert, nontoxic, and nonflammable. Because CFCs are chemically stable, once emitted, they remain in the atmosphere for hundreds or thousands of years. Because they are not found in nature, these molecules absorb reflected infrared radiation at wavelengths that otherwise would be largely unabsorbed, and they are potent greenhouse gases, with direct global warming potentials hundreds or thousands of times greater, grampergram, than that of carbon dioxide. Because their concentrations in the atmosphere are relatively small, however, their current levels of radiative forcing are low. (See page 10 for a discussion of global warming and radiative forcing.)

Because of their long atmospheric lives, a portion of the CFCs emitted into the atmosphere eventually find their way into the stratosphere, where they can be destroyed by sunlight. This reaction, however, releases free chlorine atoms into the stratosphere, and the free chlorine atoms tend to combine with stratospheric ozone, which protects the surface of the Earth from certain wavelengths of potentially damaging solar ultraviolet radiation (ultraviolet radiation, for example, causes human and animal skin cancers).

The threat posed by CFCs to the ozone layer has caused the United States and many other countries to commit themselves to phasing out the production of CFCs and their chemical cousins, hydrochlorofluorocarbons (HCFCs), pursuant to an international treaty, the 1987 Montreal Protocol. As use of CFCs has declined, many related chemicals have emerged as alternatives, including HCFCs and HFCs. HCFCs are similar to CFCs, but they are more reactive and consequently have shorter atmospheric lives, with less effect on the ozone layer and smaller direct global warming effects. HCFCs are also being phased out, but over a longer time scale. The ozonedepleting substances with the most potential to influence climate—CFC11, CFC12, and CFC113—are beginning to show reduced growth rates in atmospheric concentrations in the aftermath of the Montreal Protocol. The present radiative forcing of CFC11 is about 0.065 watts per square meter, and that of CFC12 is around 0.2 watts per square meter.[24]

HFCs have no chlorine and consequently have no effect on the ozone layer, but they are powerful greenhouse gases. The three most prominent HFCs in the atmosphere today are HFC23, HFC134a, and HFC152a.

HFC23 is formed as a byproduct of HCFC22 production, which is being phased out under the Montreal Protocol. Although HFC23 is very longlived (260 years), the growth rate in its atmospheric concentration has begun to level off in accordance with reductions in HCFC22 production. HFC134a production was rare before 1990, but in 1994 HFC134a was adopted as the standard motor vehicle air conditioning refrigerant in virtually all new cars made in America. HFC134a has a lifetime of 13.8 years, and emissions have grown rapidly from near zero in 1990 to 0.034 MMT in 2002.[25] HFC152a emissions have risen steadily since about 1995, but its short lifetime of 1.4 years has kept concentration levels below 1 part per trillion.

Another new class of engineered halocarbons is the PFCs, which include perfluoromethane (CF_4) and perfluoroethane (C_2F_6). PFCs are emitted as byproducts of aluminum smelting and are increasingly being used in the manufacture of semiconductors. They are powerful greenhouse gases and extremely longlived. Perfluoromethane has a 100year global warming potential (GWP) of 5,700 and a lifetime in excess of 50,000 years. Perfluoroethane has a GWP of 11,900 and a lifetime of 10,000 years. Perfluoromethane is a naturally occurring compound in fluorites, and emissions from this source create a natural abundance of 40 parts per trillion in the atmosphere. Increases in anthropogenic emissions, growing at about 1.3 percent annually, have raised atmospheric concentrations to 80 parts per trillion.[26] Perfluoroethane does not occur naturally in the atmosphere, and current concentrations (3.0 parts per trillion) are attributable to anthropogenic emissions, which are growing by 3.2 percent annually. Sinks for PFCs are photolysis and ion reactions in the mesosphere.[27]

Sulfur hexafluoride is used as an insulator in largescale electrical equipment and as a cover gas in magnesium smelting. It is not a halocarbon, but it is a powerful greenhouse gas. SF6 has a 100year GWP of 22,200 and a lifetime of 3,200 years. Like perfluoromethane, SF6 occurs naturally in fluorites, which produce a natural abundance of 0.01 parts per trillion in the atmosphere. Current atmospheric concentrations (3.0 parts per trillion) can be traced to anthropogenic emissions, which grew by approximately 7 percent annually during the 1980s and 1990s. Also like PFCs, sinks for SF6 are photolysis and ion reactions in the mesosphere.

There may be other chemicals not yet identified that exhibit radiative properties similar to those of the halocarbons and other gases described above. One recent discovery identified trifluoromethyl sulfur pentafluoride (SF_5CF_3) as a new anthropogenic greenhouse gas in the atmosphere.[28] It is believed that SF_5CF_3 is created by the breakdown of SF6 in highvoltage equipment, which produces CF3 that reacts with SF5 radicals resulting from highvoltage discharges. Its atmospheric concentration has grown from near zero in 1960 to 0.12 parts per trillion in 1999. To date, SF_5CF_3 has the largest radiative forcing on a permolecule basis of any gas found in the atmosphere.[29] The UNFCCC does not yet specifically address this gas.

A number of chemical solvents are also strong greenhouse gases. The solvents carbon tetrachloride (GWP of 1,800 and lifetime of 35 years) and methyl chloroform (GWP of 140 and lifetime of 4.8 years), however, are regulated in the United States for the purposes of both ozone depletion and toxicity. All these gases have direct radiative forcing effects, which are offset to some degree by their ozonedepleting effects.

With the advent of the United Nations Framework Convention and the Kyoto Protocol, the halocarbon and other industrial chemicals can be grouped into two categories:

- Ozonedepleting chemicals regulated under the Montreal Protocol but excluded from the Framework Convention (CFCs, HCFCs, and others)
- "Kyoto gases" (HFCs, PFCs, and SF_6).

The "Kyoto gases" are deemed to "count" for the purposes of meeting national obligations under the Framework Convention; however, ozonedepleting chemicals regulated by the Montreal Protocol are excluded.

Other Important Radiative Gases. There are a number of additional gases and particles, resulting in part from human sources, that produce radiative forcing of the Earth's climate but are not included under the Framework Convention or the Montreal Protocol. In general, these gases are shortlived, they have only indirect climate effects, or there is a fair amount of uncertainty about their climatic impacts. They can be broken down into three general classes: (1) ozone, both tropospheric and stratospheric; (2) criteria pollutants that are indirect greenhouse gases; and (3) aerosols, including sulfates and black soot.

Ozone (O_3) is present in both the troposphere and the stratosphere. Tropospheric ozone is not directly emitted into the atmosphere but instead forms through the photochemical reactions of various ozone precursors (primarily, nitrogen oxides and volatile organic compounds). In the troposphere, ozone acts as a direct greenhouse gas. The lifetime of ozone in the atmosphere varies from weeks to months, which imparts an element of uncertainty in estimating tropospheric ozone's radiative forcing effects. The IPCC estimates that the radiative forcing of tropospheric ozone is 0.35 ± 0.2 watts per square meter.[30] The depletion of stratospheric ozone due to the emission of halocarbons, on the other hand, has tended to cool the planet. The IPCC estimates that the cooling due to stratospheric ozone depletion is on the order of 0.15 ± 0.1 watts per square meter.[31] As the ozone layer recovers, however, due to the impacts of the Montreal Protocol, it is expected that stratospheric ozone will exert a positive radiative forcing effect on the Earth's climate.

Carbon monoxide, nitrogen oxides, and volatile organic compounds are indirect greenhouse gases. Regulated in the United States pursuant to the Clean Air Act, they are often referred to as "criteria pollutants." They are emitted primarily as byproducts of combustion (both of fossil fuels and of biomass), and they influence climate indirectly through the formation of ozone and their effects on the lifetime of methane emissions in the atmosphere. Carbon monoxide, through its effects on hydroxyl radicals, can help promote the abundance of methane in the atmosphere, as well as increase ozone formation. Some IPCC model calculations indicate that 100 metric tons of carbon monoxide emissions is equivalent to emissions of about 5 metric tons of methane.[32]

Nitrogen oxides, including NO and NO_2, influence climate by their impacts on other greenhouse gases. Nitrogen oxides not only promote ozone formation, they also impact (negatively) methane and HFC concentrations in the atmosphere. The deposition of nitrogen oxides could also reduce atmospheric carbon dioxide concentrations by fertilizing the biosphere.[33]

Volatile organic compounds (VOCs), although they have some shortlived direct radiativeforcing properties, primarily influence climate indirectly via their promotion of ozone formation and production of organic aerosols. The main sources of global VOC emissions are vegetation (primarily tropical) (1,382 $MMTCO_2e$), fossil fuels (590 $MMTCO_2e$), and biomass burning (121 $MMTCO_2e$).[34]

Aerosols, which are small airborne particles or droplets, also affect the Earth's climate. Aerosols have both direct effects, through their ability to absorb and scatter solar and thermal radiation, and indirect effects, through their ability to modify the physical properties and amount of clouds. In terms of their potential impacts on climate, the most prominent aerosols are sulfates, fossil fuel black carbon aerosols (black soot), fossil fuel organic carbon aerosols, and biomassburning aerosols.

One of the primary precursors of sulfates is sulfur dioxide (SO_2), which is emitted largely as a byproduct from the combustion of sulfurcontaining fossil fuels, particularly coal. Sulfur dioxide reacts in the air to form sulfate compounds. The major source of anthropogenic black soot and organic carbon aerosols is the burning of fossil fuels, primarily coal and diesel fuels. Biomassburning aerosols are formed by the incomplete combustion of forest products. The IPCC estimates the direct radiative forcing for aerosols as follows: sulfates, 0.4 watts per square meter; black soot, +0.2 watts per square meter; fossil fuel organic carbon, 0.1 watts per square meter; and biomassburning aerosols, 0.2 watts per square meter.[35] Although the indirect climate effects of aerosols are uncertain, some preliminary evidence points to an indirect cooling effect due to cloud formation.[36]

Relative Forcing Effects of Various Gases

The ability of a greenhouse gas to affect global temperatures depends not only on its radiative or heattrapping properties but also on its lifetime or stability in the atmosphere. Because the radiative properties and lifetimes of greenhouse gases vary greatly, comparable increases in the concentrations of different greenhouse gases can have vastly different heattrapping effects. The cumulative effect (radiative forcing—measured in watts per square meter) can vary substantially from the marginal impact of a gas. For example, among the "Kyoto gases," carbon dioxide is the most prominent in terms of emissions, atmospheric concentration, and radiative forcing (1.46 watts per square meter), but it is among the least effective as a greenhouse gas in terms of the marginal impact of each additional gram of gas added to the atmosphere. Other compounds, on a grampergram basis, appear to have much greater marginal effects.

There has been extensive study of the relative effectiveness of various greenhouse gases in trapping the Earth's heat. Such research has led to the development of the concept of a "global warming potential," or GWP. The GWP is intended to illustrate the relative impacts on global warming of an additional unit of a given gas relative to carbon dioxide over a specific time horizon. The IPCC has conducted an extensive research program aimed at summarizing the effects of various greenhouse gases through a set of GWPs. The results of that work were originally released in 1995 in an IPCC report, *Climate Change 1994,*[37] and subsequently updated in *Climate Change 1995*[38] and *Climate Change 2001*.[39] The box on page 12 provides details on the differences in emission calculations using the GWP values from the two assessments.

The calculation of a GWP is based on the radiative efficiency (heatabsorbing ability) of the gas relative to the radiative efficiency of the reference gas (carbon dioxide), as well as the removal process (or decay rate) for the gas relative to the reference gas over a specified time horizon. The IPCC, however, has pointed out that there are elements of uncertainty in calculating GWPs.[40] The uncertainty takes several forms:

- The radiative efficiencies of greenhouse gases do not necessarily stay constant over time (as calculated in GWPs), particularly if the abundance of a gas in the atmosphere increases. Each gas absorbs radiation in a particular set of wavelengths, or "window," in the spectrum. In some cases, where concentrations of the gas are low and no other gases block radiation in the same window, small emissions of the gas will have a disproportionate absorptive effect. However, if concentrations of the gas rise over time, a larger and larger portion of the total light passing through the "window" will already have been captured, and the marginal effects of additional emissions will not be as large. Therefore, the effect of an additional unit of emission of a gas that is relatively plentiful in the atmosphere, such as water vapor or carbon dioxide, tends to be less than that of a rare gas, such as sulfur hexafluoride. This "diminishing return" effect implies that increasing the concentration of a particular gas reduces the impact of additional quantities of that gas. Thus, the relative impacts of various gases will change as their relative concentrations in the atmosphere change.
- The residence time of greenhouse gases in the atmosphere (particularly, carbon dioxide) are uncertain. Various natural processes cause many greenhouse gases to decompose into other gases or to be absorbed by the ocean or ground. The amount of time it takes for natural processes to remove a unit of emissions from the atmosphere is often referred to as the "atmospheric lifetime." Some gases, such as CFCs, have very long atmospheric lifetimes, in the hundreds of years. Others, such as carbon monoxide, have lives measured in hours or days. Methane, which decays into carbon dioxide over a period of a few years, has a much larger shortrun effect on global warming than does an equivalent amount of carbon dioxide; however, over longer and longer periods—from 10 years to 100 years to 500 years, for example—the differences between the GWPs of methane and carbon dioxide become less significant, because carbon dioxide has a longer atmospheric lifetime than methane.

Table 4 summarizes the consensus results of the most recent studies by scientists working on behalf of the IPCC, showing estimates of atmospheric lifetimes and global warming potentials across various time scales. For the purposes of calculating "CO_2 equivalent" units for this report, 100year GWPs are used.

The GWPs discussed above are direct GWPs in that they consider only the direct impact of the emitted gas. The IPCC has also devoted effort to the study of indirect GWPs. Indirect GWPs are based on the climatic impacts of the atmospheric decomposition of a gas into other gases. A number of gases—including methane, carbon monoxide, halocarbons, and nitrogen oxides—are thought to have indirect climatic effects. Methane indirectly influences the climate through ozone formation and the production of carbon dioxide. Carbon monoxide can promote ozone formation and extend the lifetime of methane in the atmosphere, which results in a positive indirect GWP. Some CFCs and HCFCs produce an indirect cooling effect by removing ozone from the stratosphere. The indirect cooling effect leads to lower net GWPs in a number of cases, but in most cases their net GWPs are still positive. Nitrogen oxides promote the formation of tropospheric ozone and, thus, have a positive indirect GWP—on the order of 5 for surface emissions and 450 for aircraft emissions.[41]

Comparison of Global Warming Potentials from the IPCC's Second and Third Assessment Reports

Global warming potentials (GWPs) are used to compare the abilities of different greenhouse gases to trap heat in the atmosphere. GWPs are based on the radiative efficiency (heatabsorbing ability) of each gas relative to that of carbon dioxide (CO_2), as well as the decay rate of each gas (the amount removed from the atmosphere over a given number of years) relative to that of CO_2. The GWP provides a construct for converting emissions of various gases into a common measure, which allows climate analysts to aggregate the radiative impacts of various greenhouse gases into a uniform measure denominated in carbon or carbon dioxide equivalents. The table at the right compares the GWPs published in the Second and Third Assessment Reports of the Intergovernmental Panel on Climate Change (IPCC).

In compiling its greenhouse gas emissions estimates, the Energy Information Administration attempts to employ the most current data sources. For that reason, GWPvalues from the IPCC's Third Assessment Report are used in this report. It is important to point out, however, that countries reporting to the United Nations Framework Convention on Climate Change (UNFCCC), including the United States, have been basing their estimates on GWPs from the IPCC's Second Assessment Report. The UNFCCC Guidelines on Reporting and Review, adopted before the publication of the Third Assessment Report, require emission estimates to be based on the GWPs in the IPCC's Second Assessment Report. This will probably continue in the short term, until the UNFCCC reporting rules are changed. Following the current rules, the U.S. Environmental Protection Agency, which compiles the official U.S. emissions inventory for submission to the UNFCCC, presented estimates based on the GWPs published in the Second Assessment Report in its report, Inventory of U.S. Greenhouse Gas Emissions and Sinks: 19902003, released in April 2005.

The table below shows U.S. carbon dioxide equivalent greenhouse gas emissions calculated using the IPCC's 1996 (Second Assessment Report) and 2001 (Third Assessment Report) GWPs. The estimate for total U.S. emissions in 2004 is 0.7 percent higher when the revised GWPs are used. The estimates for earlier years generally follow the same pattern.

Therefore, trends in growth of greenhouse gas emissions are similar for both sets of GWP values. Using the 2001 GWPs, estimates of carbon dioxide equivalent methane emissions are 9.5 percent higher, and carbonequivalent nitrous oxide emissions are 4.5 percent lower. Carbon dioxide equivalent emissions of HFCs, PFCs, and SF_6 are lower for some years and higher for others, depending on the relative shares of the three gases.

Comparison of 100Year GWP Estimates from the IPCC's Second (1996) and Third (2001) Assessment Reports

Gas	1996 IPCC GWP	2001 IPCC GWP
Methane.	21	23
Nitrous Oxide.	310	296
HFC-23	11,700	12,000
HFC-125	2,800	3,400
HFC-134a	1,300	1,300
HFC-143a	3,800	4,300
HFC-152a	140	120
HFC-227ea	2,900	3,500
HFC-236fa.	6,300	9,400
Perfluoromethane (CF_4). . . .	6,500	5,700
Perfluoroethane (C_2F_6).	9,200	11,900
Sulfur Hexafluoride (SF_6). . .	23,900	22,200

Sources: UNFCCC, Second Assessment Report (1996) and Third Assessment Report (2001).

Gas	IPCC GWP		Annual GWP-Weighted Emissions (Million Metric Tons Carbon Dioxide Equivalent)								
			1990			2003			2004		
	1996	2001	1996 GWP	2001 GWP	Percent Difference	1996 GWP	2001 GWP	Percent Difference	1996 GWP	2001 GWP	Percent Difference
Carbon Dioxide.	1	1	5,002	5,002	0.0	5,872	5,871	0.0	5,973	5,973	0.0
Methane	21	23	659	721	9.5	579	634	9.5	584	639	9.5
Nitrous Oxide	310	296	353	337	-4.5	351	335	-4.5	370	354	-4.5
HFCs, PFCs, and SF_6. .	M	M	92	88	-4.0	137	142	4.2	148	156	5.3
Total.	—	—	**6,106**	**6,149**	**0.7**	**6,938**	**6,983**	**0.7**	**7,075**	**7,112**	**0.7**

M = mixture of gases.

Sources: Estimates provided in this report; and UNFCCC, Second Assessment Report (1996) and Third Assessment Report (2001).

Current U.S. Climate Change Initiatives

The Bush Administration is pursuing a broad range of strategies to address the issues of global climate change through the implementation of multiple new initiatives. Details of these initiatives were initially provided on February 14, 2002, when the President announced the Global Climate Change Initiative. This initiative sets a national goal for the United States to reduce its greenhouse gas intensity (total greenhouse gas emissions per unit of gross domestic product [GDP]) by 18 percent between 2002 and 2012 through voluntary measures (see box on page 14).

To meet this goal and encourage the development of strategies and technologies that can be used to limit greenhouse gas emissions both at home and abroad, the Administration has implemented a number of related initiatives, including the following:[42]

- **Climate Change Technology Program (CCTP):** The CCTP is a multiagency program to accelerate the development and deployment of key technologies that can achieve substantial reductions in greenhouse gas emissions. CCTP includes climate changerelated technology research, development, and deployment efforts as well as voluntary programs.
- **Climate Change Science Program (CCSP):** The CCSP is a Federal, multiagency research program to investigate natural and humaninduced changes in the Earth s global environmental system; to monitor, understand, and predict global change; and to provide a sound scientific basis for national and international decisionmaking.
- **International Cooperation:** The United States is engaged in international efforts on climate change, both through multilateral and bilateral activities. Multilaterally, the United States is the largest donor to activities under the UNFCCC and the IPCC. Since 2001, the United States has launched bilateral partnerships with numerous countries on issues ranging from climate change science, to energy and sequestration technologies, to policy approaches. As an example, a new international effort is the AsiaPacific Partnership on Clean Development, which involves the United States, Australia, China, India, Japan, and South Korea. The partnership focuses on voluntary measures aimed at creating new investment opportunities to build cleaner, more efficient capacity in energy generation and related systems, including methane capture and use, rural energy systems, clean coal, and civilian nuclear power, as well as advanced transportation and renewable energy systems.
- **NearTerm Greenhouse Gas Reduction Initiatives:** The Federal Government administers a wide array of voluntary, regulatory, and incentivebased programs on energy efficiency, agricultural practices, and greenhouse gas reductions. Major initiatives announced by the Bush Administration include:
  - **Climate VISION Partnership:** In February 2003, President Bush announced that 12 major industrial sectors and the membership of the Business Roundtable had committed to work with the EPA and three Federal departments (Energy, Transportation, and Agriculture) to reduce greenhouse gas emissions in the next decade. Participating industries include electric utilities; petroleum refiners and natural gas producers; automobile, iron and steel, chemical, and magnesium manufacturers; forest and paper producers; railroads; and the cement, mining, aluminum, lime, and semiconductor industries. In May 2005, the Industrial Minerals Association–North America joined the list of participating industries.
  - **Climate Leaders:** Climate Leaders is a voluntary partnership that encourages companies to establish and meet clearly defined targets for greenhouse gas emission reductions. The EPA established Climate Leaders in February 2002 and has recruited 74 partners, 38 of which have established greenhouse gas reduction goals. Climate Leaders partners include companies such as 3M and Alcoa (manufacturing); Baxter International, Johnson & Johnson, and Pfizer, (health services/ pharmaceuticals); General Motors, Volvo Trucks North America, and Mack Trucks (automotive); and American Electric Power, Entergy Corporation, and FPL Group (utilities). By joining Climate Leaders, the partners commit themselves to document their emissions of the six major greenhouse gases (carbon dioxide, methane, nitrous oxide, HFCs, PFCs, and SF_6) on a

companywide basis (including, at a minimum, all their domestic facilities). Partners are required to develop an Inventory Management Plan (IMP) and report their annual corporatelevel emissions by emission source type to the EPA, using the EPA's Annual GHG Inventory Summary and Goal Tracking Form.[43] In October 2004, the EPA issued updated guidance for corporate greenhouse gas inventories, based on an existing protocol developed by the World Resources Institute (WRI) and the World Business Council for Sustainable Development (WBCSD). The EPA has finalized guidance covering design principles and crosssector core guidance covering direct emissions from stationary combustion, indirect emissions from sales and purchases of electricity and steam, direct emissions from mobile combustion sources, direct emissions from municipal solid waste landfilling, and direct emissions of HFCs and PFCs from use of refrigeration and air conditioning equipment. The EPA has also completed draft sectorspecific guidance for core emissions from the following industries: cement, manufacture of refrigeration and air conditioning equipment (HFC and PFC emissions), and iron and steel. The EPA is currently developing sectorspecific guidance for aluminum production, pulp and paper production, semiconductor manufacturing, and SF6 from electricity distribution.

- **Enhanced 1605(b) Voluntary Emissions Reduction Registry:** Pursuant to a key objective of the Global Climate Change Initiative, announced on February 14, 2002, by President George W. Bush, the U.S. Department of Energy (DOE) is working to improve and expand the existing 1605(b) Voluntary Reporting of Greenhouse Gases Program. The primary goal of the DOE effort is to create a credible and transparent program for reporting of real reductions that support the national greenhouse gas intensity goal laid out in the Global Climate Change Initiative. An additional goal of the enhanced 1605(b) program is to allow businesses and individuals to record their reductions and ensure that reporters are not penalized under any future climate policy. The objective of improving the registry is to help motivate firms to take costeffective, voluntary actions to reduce greenhouse gas emissions, which would, in part, aid in the achievement of the Initiative's greenhouse gas intensity goal. An interagency working group has undertaken several actions to improve the Voluntary Reporting Program, including outreach efforts, solicitation of public comments, and review of the existing program. On July 8, 2002, the Secretary of Energy, joined by the Secretary of Commerce, the Secretary of Agriculture, and the EPA Administrator, submitted recommendations to the White House to guide the process for improving and expanding the Voluntary Reporting Program. In November 2003, DOE released proposed revisions to the General Guidelines, which outline the principles that would govern the revised program. A public workshop on the subject was held in Washington, DC, on January 12,2004, and DOE continued to collaborate with the USDA, the EPA, and other Federal agencies throughout 2004 in developing revised Guidelines for the Voluntary Reporting of Greenhouse Gases Program and "draft" Technical Guidelines that will specify the methods and factors to be used in measuring and estimating greenhouse gas emissions, emission reductions, and carbon sequestration under the revised program. On March 24, 2005, the Interim Final General Guidelines and a "Notice of Availability" for the Draft Technical Guidelines were published in the *Federal Register* for 60 days of public comment. Both DOE and the USDA have held workshops to solicit comments on the Interim Final General Guidelines and the Draft Technical Guidelines for the Voluntary Reporting of Greenhouse Gases Program. The DOE workshop, held on April 2627, 2005, in Arlington, Virginia, was attended by more than 150 individuals, most of whom were participants in the existing 1605(b) program.[44] The USDA workshop was held on May 5, 2005, in Riverdale, Maryland, to address technical and methodological issues associated with preparing estimates of greenhouse gas emissions and carbon sequestration from agriculture and forestry activities and reporting those emissions under DOE's revised 1605(b) program. USDA invited four organizations to perform technical evaluations of the forestry and agriculture sections of the draft technical guidelines. Evaluators were asked to assess the accuracy, complexity, clarity, consistency, and comprehensiveness of the guidelines as applied to their operations. USDA also requested views on how to streamline and improve the usability of the guidelines.[45] In a *Federal Register* notice published on May 9, 2005, DOE announced that it had extended the period for public comment on the Interim Final General Guidelines and Draft Technical Guidelines for the Voluntary Reporting of Greenhouse Gases Program to June 22, 2005, with a scheduled "effective date" of September 20, 2005. As this report goes to press, however, DOE still is in the process of making a determination as to whether addressing the comments received will require an extension of the scheduled "effective date" of the revised guidelines beyond September 20, 2005.[46]

Table 4. Numerical Estimates of Global Warming Potentials Compared With Carbon Dioxide (Kilogram of Gas per Kilogram of Carbon Dioxide)

Gas	Lifetime (Years)	Direct Effect for Time Horizons of		
		20 Years	100 Years	500 Years
Carbon Dioxide	5 – 200[a]	1	1	1
Methane	12	62	23	7
Nitrous Oxide	114	275	296	156
HFCs, PFCs, and Sulfur Hexafluoride				
HFC-23	260	9,400	12,000	10,000
HFC-125	29	5,900	3,400	1,100
HFC-134a	13.8	3,300	1,300	400
HFC-152a	1.4	410	120	37
HFC-227ea	33	5,600	3,500	1,100
Perfluoromethane (CF_4)	50,000	3,900	5,700	8,900
Perfluoroethane (C_2F_6)	10,000	8,000	11,900	18,000
Sulfur Hexafluoride (SF_6)	3,200	15,100	22,200	32,400

[a]No single lifetime can be defined for carbon dioxide due to different rates of uptake by different removal processes.

Note: The typical uncertainty for global warming potentials is estimated by the Intergovernmental Panel on Climate Change at ±35 percent.

Source: Intergovernmental Panel on Climate Change, *Climate Change 2001: The Scientific Basis* (Cambridge, UK: Cambridge University Press, 2001), pp. 38 and 388389.

Trends in U.S. Carbon Intensity and Total Greenhouse Gas Intensity

From 2003 to 2004, the greenhouse gas intensity of the U.S. economy fell from 677 to 662 metric tons per million 2000 dollars of GDP (2.1 percent), continuing a trend of decreases in both carbon intensity (see figure at right) and total greenhouse gas intensity. As shown in the table below, declines in carbon intensity by decade have ranged from a low of 3.3 percent in the 1960s to 25.6 percent in the 1980s. From 1990 to 2004, total U.S. greenhouse gas intensity fell by 23.5 percent, at an average rate of 1.9 percent per year.

The carbon intensity of the economy can largely be decomposed into two basic elements: (1) energy intensity, defined as the amount of energy consumed per dollar of economic activity; and (2) carbon intensity of energy supply, defined as the amount of carbon emitted per unit of energy. As illustrated by the formulas below, the multiplication of the two elements produces a numerical value for U.S. carbon intensity, defined as the amount of carbon dioxide emitted per dollar of economic activity:

Historical Growth Rates for U.S. Carbon Intensity

	Overall Change in Intensity (Percent)		Average Annual Change in Intensity (Percent)	
Decade	Carbon Dioxide	Total GHG	Carbon Dioxide	Total GHG
History				
1950-1960	-12.9	—	-1.4	—
1960-1970	-3.3	—	-0.3	—
1970-1980	-17.7	—	-1.9	—
1980-1990	-25.6	—	-2.7	—
1990-2000	-15.3	-17.9	-1.6	-1.9

Source: Energy Information Administration, Annual Energy Review 2004, DOE/EIA0384(2003) (Washington, DC, August 2005), and estimates presented in Appendix B of this report.

Energy Intensity x *Carbon Intensity of Energy Supply* = *Carbon Intensity of the Economy*

or, algebraically,

(Energy/GDP) x *(Carbon Emissions/Energy)* = *(Carbon Emissions/GDP)*

Components of Energy Intensity. Since World War II the U.S. economy has been moving away from traditional "smokestack" industries toward more servicebased or informationbased enterprises. This has meant that over the second half of the 20th century economic growth was less tied to growth in energy demand than it was during the period of industrialization in the 19th and early 20th century. Other factors contributing to decreases in energy intensity include:

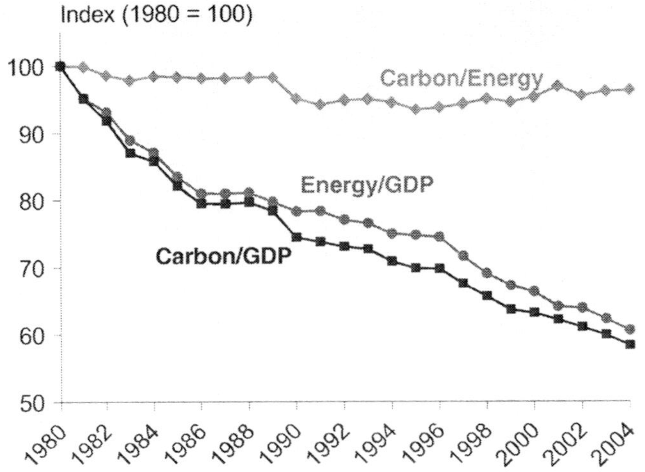

Source: Estimates presented in Appendix B of this report.

Intensity Ratios: Carbon/GDP, Carbon/Energy, and Energy/GDP

- Improvements in the energy efficiency of industrial equipment as new materials and methods improved performance in terms of energy inputs versus outputs.
- Increased efficiency of transportation equipment as lighter materials and more efficient engines entered the marketplace.
- Improvements in commercial and residential lighting, refrigeration, and heating and cooling equipment.
- Developments in new electricity generating technologies, such as combinedcycle turbines.

Further reductions in energy intensity, which are projected to continue, will among other things promote deeper reductions in U.S. carbon intensity.

Components of the Carbon Intensity of Energy Supply. Changes in the carbon intensity of energy supply have been less dramatic than changes in energy intensity. There was a slow but steady decline from 1980 until about the mid1990s, after which it has remained relatively unchanged. The primary reason for the decline has been the development of nuclear power, which is carbonfree and therefore weights the fuel mix toward lower carbon intensity. Other factors that can decrease the carbon intensity of the energy supply include:

- Development of new renewable resources, such as wind power, for electricity generation
- Substitution of natural gas for coal and oil in power generation
- Transportation fuels with a higher biogenic component, such as ethanol.

International Developments in Global Climate Change

The primary international agreement addressing climate change is the UNFCCC, which opened for signature at the "Earth Summit" in Rio de Janeiro, Brazil, in June 1992 and entered into force in March 1994.[47] The agreement currently has 185 signatories, including the United States. The objective of the Framework Convention is stated as follows:

> *The ultimate objective of this Convention and any related legal instruments that the Conference of the Parties may adopt is to achieve, in accordance with the relevant provisions of the Convention, stabilization of greenhouse gas concentrations in the atmosphere at a level that would prevent dangerous ant hropogenic interference with the climate system.[48]*

The Framework Convention divided its signatories into three groups: the countries listed in Annex I; Annex II, which comprises the Annex I countries minus the countries with economies in transition; and nonAnnex I countries, which include countries that ratified or acceded to the UNFCCC but are not included in Annex I. The Annex I countries include the 24 original members of the Organization for Economic Cooperation and Development (OECD) (including the United States), the European Union, and 14 countries with economies in transition (Russia, Ukraine, and Eastern Europe).[49]

The Convention requires all parties to undertake "policies and measures" to limit emissions of greenhouse gases, and to provide national inventories of emissions of greenhouse gases (Article 4.1a and b). Annex I parties are further required to take actions "with the aim of returning... to their 1990 levels these anthropogenic emissions of carbon dioxide and other greenhouse gases" (Article 4.2a and b). The signatories subsequently agreed that Annex I parties should provide annual inventories of greenhouse gas emissions.

The Kyoto Protocol

The Kyoto Protocol to the UNFCCC, negotiated in December 1997, is a set of quantified greenhouse gas emissions targets for Annex I countries, which collectively are about 5 percent lower than the 1990 emissions of those countries taken as a group.[50] Developing country signatories do not have quantified targets.[51] The conditions for ratification of the Kyoto Protocol were met in November 2004, following formal acceptance by the Russian Parliament and President Putin's signing of the ratifying legislation. Those actions brought the number of signatory countries to 118, with Annex I countries representing 61.2 percent of total Annex I carbon dioxide emissions in 1990. The Protocol entered into force in February 2005. While the United States remains a participant in the Framework Convention, it is not a participant in the Kyoto Protocol.

Recent and Upcoming Conferences of the Parties and Other International Events

Since the negotiation of the Kyoto Protocol in 1997, much of the work done at periodic (usually annual) meetings of the UNFCCC Conference of the Parties (COP) has been focused on filling in details related to the operation of the UNFCCC, the Protocol, and their respective mechanisms.

COP10

COP10, held in Buenos Aires, Argentina, from December 6 through December 17, 2004, marked the 10th anniversary of the entry into force of the UNFCCC, which was a central theme for the meeting.[52] In addition to the accomplishments of the first 10 years of the Convention and future challenges, discussions at COP10 highlighted a range of climaterelated issues, including the impacts of climate change and adaptation measures, mitigation policies and their impacts, and technology. Participants also considered the entry into force of the Kyoto Protocol, which was enabled by Russia's ratification.

COP11 and MOP1

Canada recently hosted the first Meeting of the Parties to the Kyoto Protocol (MOP1) in conjunction with the eleventh meeting of the Conference of Parties to the Framework Convention (COP11). The meetings were held in Montreal, Canada, from November 28 to December 9, 2005.

G8 Summit in Gleneagles, Scotland

In a communiqué on climate change and the related topics of clean energy and sustainable development, the leaders of the G8 who met July 68, 2005, in Gleneagles, Scotland, outlined several key points that summarize their position:

- They declared that, "climate change is a serious and longterm challenge that has the potential to affect every part of the globe."
- They acknowledged that "... use of energy from fossil fuels, and other human activities, contribute in large part to increases in greenhouse gases associated with the warming of our Earth's surface."
- The communiqué further stated that, "While uncertainties remain in our understanding of climate science, we know enough to act now to put ourselves on a path to slow and, as the science justifies, stop and then reverse the growth of greenhouse gases."

Other points further stressed the relationship between climate change, clean energy, and sustainable development. A Dialogue on Climate Change, Clean Energy and Sustainable Development was proposed, which would:

- Address the strategic challenge of transforming our energy systems to create a more secure and sustainable future
- Monitor implementation of the commitments made in the Gleneagles Plan of Action
- Share best practice between participating governments
- Produce a report due at the 2008 G8 Summit, to be hosted by Japan.

The leaders also reaffirmed their commitment to the UNFCCC and the importance of the IPCC. They emphasized the UNFCCC objective of stabilizing atmospheric concentrations of greenhouse gases at a level that "prevents dangerous anthropogenic interference with the climate system." They agreed to support efforts at COP11 to "move forward in that forum the global discussion on longterm cooperative action to address climate change."

The importance of sustainable development in emerging economies was stressed, and to that end the leaders of Brazil, China, India, Mexico, and South Africa also participated in the Summit. Collectively they called for stronger efforts by developed countries to reduce emissions and agreed to provide financial and technical assistance to developing countries. Additionally, the World Bank and the International Energy Agency were highlighted for their potentially important roles in the process of clean and sustainable development.

The Gleneagles Plan of Action adopted by the G8 leaders identifies a range of activities to promote research, information exchange, and cooperation on energy efficiency, renewable and other clean energy sources, and adaptation to climate change.

Key elements of the Plan include:

- A review of building codes and appliance and vehicle standards to identify best practices and opportunities for coordination
- An extension of the use of labeling on vehicles and appliances to raise consumer awareness of energy consumption
- Encouragement of multilateral development banks to expand the use of voluntary energy savings assessments of proposed investments in energyintensive sectors; to explore opportunities to increase investments in renewable energy and energy efficiency technologies; and to work with borrower countries to identify less greenhouse gasintensive growth options.[53]

Carbon Dioxide Emissions

Overview

U.S. Anthropogenic Carbon Dioxide Emissions, 1990-2004

	Carbon Dioxide	Carbon Equivalent
Estimated 2004 Emissions (Million Metric Tons)	5,973.0	1,629.0
Change Compared to 2003 (Million Metric Tons)	101.2	27.6
Change from 2003 *(Percent)*	*1.7%*	*1.7%*
Change Compared to 1990 (Million Metric Tons)	970.7	264.7
Change from 1990 *(Percent)*	*19.4%*	*19.4%*
Average Annual Increase, 1990-2004 *(Percent)*	*1.3%*	*1.3%*

Carbon dioxide emissions in the United States and its Territories were 5,973.0 million metric tons (MMT) in 2004, 101.2 MMT (1.7 percent) more than in 2003 (Table 5). The increase in emissions from 2003 to 2004 can be attributed mainly to an increase in overall U.S. economic growth in 2004 of 4.2 percent, including an increase in industrial production. As a result of the higher economic growth rate, fossil fuel consumption and carbon dioxide emissions in the transportation and industrial sectors increased (Table 6). The 1.7percent increase in total emissions from 2003 to 2004 followed an increase of 1.1 percent, or 63.3

MMT, from 2002 to 2003 (Figure 1). Since 1990, total U.S. carbon dioxide emissions have increased by an average of about 1.3 percent per year.

Carbon dioxide emissions represent about 84 percent of total U.S. greenhouse gas emissions. In the United States, most carbon dioxide (98 percent) is emitted as the result of the combustion of fossil fuels; consequently, carbon dioxide emissions and energy use are highly correlated. (The remaining 2 percent of carbon dioxide emissions comes from a variety of other industrial sources.) Historically, economic growth, the weather, the carbon and energy intensity of the economy, and movements in energy prices have caused yeartoyear fluctuations in energy consumption and resulting carbon dioxide emissions.

In both the residential and commercial sectors, 2004 energyrelated carbon dioxide emissions were virtually unchanged from 2003 levels (Table 6). In the residential sector, emissions of carbon dioxide decreased by 0.1 percent, from 1,213.2 MMT in 2003 to 1,212.0 MMT in 2004. In the commercial sector, carbon dioxide emissions increased by 0.3 percent, from 1,021.1 MMT in 2003 to 1,024.2 MMT in 2004. Growth in energy consumption, and subsequently emissions, was mitigated in both sectors, because both a warmer winter and a cooler summer than in 2003 reduced heating and cooling degreedays by 3.8 and 4.4 percent, respectively.

Industrial production rose by 4.1 percent in 2004, and industrial emissions of carbon dioxide rose by 2.6 percent, from 1,685.6 MMT in 2003 to 1,730.2 MMT in 2004 (Table 6). Trends in industrial emissions are driven in part by growth patterns in the six most energyintensive manufacturing industries, which account for about twothirds of total industrial emissions of carbon dioxide. In 2004, there was growth in all six energyintensive industries. Three of those manufacturing industries grew at rates greater than the overall GDP growth rate of 4.2 percent: paper by 5.6 percent, primary metals by 5.5 percent, and nonmetallic minerals by 4.6 percent. The other three grew by less than overall GDP: the petroleum industry by 3.9 percent, the food industry by 3.2 percent, and chemicals by 2.0 percent.

Estimates for 2004 indicate that, as a result of strong growth in demand for transportation services, carbon dioxide emissions in the transportation sector increased by 3.1 percent, from 1,875.7 MMT in 2003 to 1,933.7 MMT in 2004 (Table 6). If these preliminary numbers prove accurate, this would be the highest growth rate in transportationsector emissions of carbon dioxide during the 1990 to 2004 period.

Net generation of electricity increased by 1.9 percent from 2003 to 2004, although carbon dioxide emissions from the electric power sector increased by only 0.9 percent, from 2,278.8 MMT in 2003 to 2,298.6 MMT in 2004 (Table 6). Accordingly, the overall carbon intensity of U.S. electricity production fell by about 1.1 percent. The decrease in carbon intensity was driven by an increase in the use of natural gas to generate electricity, while coal use for generation was essentially flat. Natural gas has a lower carbon content than the coal it displaced and generally higher efficiency when used to produce electricity. In addition, generation from "noncarbon" nuclear and renewable fuels increased by 21 billion kilowatthours (2.0 percent).[54]

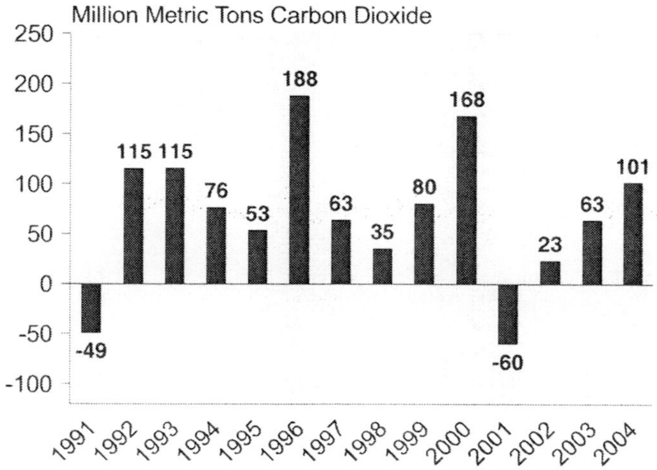

Source: Estimates presented in this chapter.

Figure 1. Annual Change in U.S. Carbon Dioxide Emissions, 1990-2004

In this report, the electric power sector is defined as all utilities, nonutilities, and combined heat and power (CHP) facilities whose primary business is the production of electric power. Carbon dioxide emissions from generators that produce electric power as part of an industrial or commercial operation—that is, businesses that produce electricity primarily for their own use—are not included in the electric power sector total but are assigned to the industrial or commercial sector according to the classification of the business. In addition, the emissions totals reported above for the energy enduse sectors (residential, commercial, industrial, and transportation) include their shares of total electric power sector emissions.

Nonfuel uses of fossil fuels, principally petroleum, both emit and sequester carbon dioxide over their life cycles. In 2004, carbon dioxide emissions from nonfuel uses of fossil fuels totaled 114.3 MMT, an 8.6percent increase from the 2003 total of 105.2 MMT (Table 5). Nonfuel uses of fossil fuels also resulted in carbon sequestration equal to 316.6 million metric tons carbon dioxide equivalent ($MMTCO_2e$) in 2004, a 7.7percent increase from the 2003 level of 294 $MMTCO_2e$.[55] The major fossil fuel products that emit and sequester carbon include liquefied petroleum gas (LPG) and feedstocks for plastics and other petrochemicals. Asphalt and road oils are a major source of sequestration, but they do not emit carbon dioxide. It is estimated that, of the amount of carbon dioxide sequestered in the form of plastic, about 11.0 MMT was emitted as carbon dioxide from the burning of the plastic components of municipal solid waste to produce electricity in 2004. The 2003 estimate of 18.8 MMT is used in this report as an estimate for total 2004 emissions of carbon dioxide from all burning of wastes, because the 2004 estimate is not yet available.

Emissions of carbon dioxide from other sources— including cement production, industrial processes, waste combustion, carbon dioxide in natural gas, and gas flaring—increased by 2.7 percent, from 102.2 MMT in 2003 to 105.0 MMT in 2004 (Table 5).

Energy Consumption

Energy EndUse Sector Sources **of U.S. Carbon Dioxide Emissions, 1990-2004**

Sector	Million Metric Tons Carbon Dioxide		*Percent Change*	
	1990	2004	1990-2004	2003-2004
Residential	953.7	1,212.0	*27.1%*	*-0.1%*
Commercial	780.7	1,024.2	*31.2%*	*0.3%*
Industrial	1,692.2	1,730.2	*2.2%*	*2.6%*
Transportation	1,569.9	1,933.7	*23.2%*	*3.1%*

Note: Electric power sector emissions are distributed across sectors.

The consumption of energy in the form of fossil fuel combustion is the largest single contributor to greenhouse gas emissions in the United States and the world. Of total 2004 U.S. carbon dioxide emissions (adjusting for U.S. Territories and bunker fuels), about 98 percent, or 5,868.0 MMT carbon dioxide, resulted from the combustion of fossil fuels. This figure represents an increase of 1.7 percent from 2003 levels.

In the short term, yeartoyear changes in energy consumption and carbon dioxide emissions tend to be dominated by weather, economic fluctuations, and movements in energy prices. Over longer time spans, changes in energy consumption and emissions are also influenced by other factors, such as population shifts and energy consumers' choice of fuels, appliances, and capital equipment (e.g., vehicles, aircraft, and industrial plant and equipment). The energyconsuming capital stock of the United States—cars and trucks, airplanes, heating and cooling plants in homes and businesses, steel mills, aluminum smelters, cement plants, and petroleum refineries—changes slowly from one year to the next, because capital stock usually is retired only when it begins to break down or becomes obsolete.

The Energy Information Administration (EIA) divides energy consumption into four general enduse categories: residential, commercial, industrial, and transportation. Emissions from electricity generators, which provide electricity to the enduse sectors, are allocated in proportion to the electricity consumed in, and losses allocated to, each sector (Table 6).

Residential Sector

At 1,212.0 MMT, residential carbon dioxide emissions represented 21 percent of U.S. energyrelated carbon dioxide emissions in 2004. The residential sector's prorated share of electric power sector emissions, 837.3 MMT, accounts for more than twothirds of all emissions in the residential sector (Table 7).[56] Natural gas accounted for 22 percent (265.5 MMT), and petroleum (mainly distillate fuel oil) represented 8.9 percent (108.0 MMT). Since 1990, residential electricityrelated emissions have grown by 2.2 percent annually. In contrast, emissions from the direct combustion of fuels, primarily natural gas, in the residential sector have grown by 0.7 percent annually since 1990.

Total carbon dioxide emissions from the residential sector decreased by 0.1 percent in 2004. Yeartoyear, residential sector emissions are strongly influenced by weather. The warmer winter in 2004, relative to 2003, was a contributor to the 2004 decrease in residential sector emissions. In addition, the 2004 summer was cooler than in 2003, leading to a drop in both heating and cooling degreedays of about 4 percent.[57]

Since 1990, the growth in carbon dioxide emissions attributable to the residential sector has averaged 1.7 percent per year. Residential sector emissions in 2004 were 258.3 MMT higher than in 1990, representing 29 percent of the total increase in unadjusted U.S. energyrelated carbon dioxide emissions since 1990. Longterm trends in residential carbon dioxide emissions are strongly influenced by demographic factors, living space attributes, and building shell and appliance efficiency choices. For example, the movement of population into warmer climates tends to increase summer air conditioning consumption and promote the use of electric heat pumps, which increases emissions from electricity use (although the increase could be offset by a reduction in emissions from heating fuel combustion). Growth in the number of households, resulting from increasing population and immigration, contributes to more residential energy consumption.

Commercial Sector

Commercial sector carbon dioxide emissions, at 1,024.2 MMT, accounted for about 17 percent of total energyrelated carbon dioxide emissions in 2004, of which 78 percent (795.4 MMT) is the sector's prorated share of electricityrelated emissions (Table 8). Natural gas contributes 16 percent and petroleum 6 percent of the sector's emissions.

Commercial sector emissions largely have their origin in the lighting, space heating, and space cooling requirements of commercial structures, such as office buildings, shopping malls, schools, hospitals, and restaurants. Lighting is a significantly more important component of energy demand in the commercial sector (approximately 20 percent of total demand in 2003) than it is in the residential sector (approximately 12 percent of total demand in 2003). Heating and cooling demand accounted for approximately 41 percent of energy demand in the residential sector in 2003, and about 20 percent in the commercial sector.[58] Thus, commercial sector emissions are affected less by the weather than residential sector emissions. In the longer run, because commercial activity is a factor of the larger economy, emissions from the commercial sector are more affected by economic trends and less affected by population growth than are emissions from the residential sector.

Emissions attributable to the commercial sector's prorated share of electricity consumption increased by 1.5 percent in 2004, and emissions from the direct combustion of fuels (dominated by natural gas, as in the residential sector) decreased by 3.6 percent. Overall, carbon dioxide emissions related to commercial sector activity increased by 0.3 percent—from 1,021.1 to 1,024.2 MMT—between 2003 and 2004 (Table 8). Since 1990, commercial emissions growth has averaged 2.0 percent per year, the largest growth of any enduse sector. Commercial sector carbon dioxide emissions have risen by 243.4 MMT since 1990, accounting for 27 percent of the total increase in U.S. unadjusted energyrelated carbon dioxide emissions.

Industrial Sector

Industrial sector emissions, at 1,730.2 MMT carbon dioxide, accounted for 29 percent of total U.S. energyrelated carbon dioxide emissions in 2004. In terms of fuel shares, electricity consumption was responsible for 38 percent of total industrial sector emissions (660.9 MMT), natural gas for 26 percent (441.9 MMT), petroleum for 26 percent (440.6 MMT), and coal for 10 percent (181.0 MMT).

Estimated 2004 energyrelated carbon dioxide emissions in the industrial sector, at 1,730.2 MMT (Table 9), were 2.6 percent higher than the 2003 emissions level of 1,685.6 MMT. Carbon dioxide emissions attributable to industrial sector energy consumption, while fluctuating from year to year, have increased by an average of only 0.2 percent per year since 1990. As a result, total energyrelated industrial emissions in 2004 were just 2.2 percent (38.0 MMT) higher than in 1990, despite a much larger economy.

A contributing factor to the slow growth in industrial sector carbon dioxide emissions (it has the lowest growth rate of any of the enduse sectors) is the erosion of the older energyintensive (and specifically coalintensive) industrial base. For example, coke plants consumed 38.9 million short tons of coal in 1990, as compared with 23.7 million short tons in 2004. Other industrial coal consumption declined from 76.3 million short tons in 1990 to 61.2 million short tons in 2004. Also, the share of manufacturing activity represented by less energyintensive industries, such as computer chip and electronic component manufacturing, has increased while the share represented by energyintensive industries has fallen.

Transportation Sector

Carbon dioxide emissions from the transportation sector, at 1,933.7 MMT, accounted for 33 percent of total U.S. energyrelated carbon dioxide emissions in 2004. Almost all (98 percent) of transportation sector emissions result from the consumption of petroleum products: motor gasoline, at 60 percent of total transportation sector emissions; middle distillates (diesel fuel) at 22 percent; jet fuel at 12 percent of the total; and residual oil (i.e., heavy fuel oil, largely for maritime use) at 2.8 percent of the sector's total emissions. Motor gasoline is used primarily in automobiles and light trucks, and middle distillates are used in heavy trucks, locomotives, and ships.

Emissions attributable to the transportation sector increased by 3.1 percent in 2004, from 1,875.7 MMT carbon dioxide in 2003 to 1,933.7 MMT in 2004 (Table 10). The fueluse patterns and related emissions sources in the transportation sector are different from those in the other enduse sectors. By far the largest single source of emissions, motor gasoline, at 1,162.6 MMT carbon dioxide, grew by 1.9 percent. Emissions from motor gasoline were mitigated somewhat by a 24percent increase in the consumption of ethanol. Carbon dioxide emissions from ethanol consumption are considered to be zero, because the carbon in the fuel is derived primarily from corn, and it is assumed that an equivalent amount of carbon will be sequestered during the corngrowing season.

Since 1990, carbon dioxide emissions related to the transportation sector have increased at an average annual rate of 1.5 percent. The growth since 1990 has meant that transportation emissions have increased by 363.7 MMT, representing 40 percent of the growth in unadjusted energyrelated carbon dioxide emissions from all sectors. Transportation is the largest contributing enduse sector to total emissions.

Electric Power Sector

Electric Power Sector Carbon Dioxide Emissions by Fuel Input, 1990-2004

	Million Metric Tons Carbon Dioxide		*Percent Change*	
Fuel	**1990**	**2004**	**1990-2004**	**2003-2004**
Petroleum	100.9	97.4	*-3.5%*	*0.3%*
Natural Gas	176.9	295.9	*67.3%*	*6.6%*
Coal	1,519.1	1,893.9	*24.7%*	*0.1%*
Total	1,803.1	2,298.6	*27.5%*	*0.9%*

The data in Table 11 represent estimates of carbon dioxide emissions for the electric power sector. These emissions when taken as a whole account for 39 percent of total U.S. energyrelated carbon dioxide emissions; in calculating sectorspecific emissions, electric power sector emissions are distributed to the enduse sectors. The electric power sector includes traditional regulated utilities, as well as independent power producers whose primary business is the generation and sale of electricity. The industrial sector and, to a much lesser extent, the commercial sector also include establishments that generate electricity; however, their primary business is not electricity generation, and so their electricityrelated emissions are included in the totals for those sectors, not in the electric power sector. Preliminary estimates indicate that carbon dioxide emissions from the electric power sector increased by 0.9 percent (19.7 MMT), from 2,278.8 MMT in 2003 to 2,298.6 MMT in 2004 (Table 11). Emissions from naturalgasfired generation increased by 6.6 percent, from coalfired generation by 0.1 percent, and from petroleumfired generation by 0.3 percent. Carbon dioxide emissions from the electric power sector have grown by 28 percent since 1990, while total unadjusted energyrelated carbon dioxide emissions have grown by 18 percent. Of the total growth in energyrelated carbon dioxide emissions from 1990 to 2004,55 percent is attributable to the electric power sector.

EnergyRelated Carbon Dioxide Emissions in Manufacturing

Manufacturing is the single largest source of energyrelated carbon dioxide emissions in the U.S. industrial sector, which also includes agriculture, forestry, fisheries, mining, and construction. The manufacturing subsector accounted for about 84 percent of energyrelated carbon dioxide emissions and 90 percent of energy consumption in the industrial sector in 2002. The table below shows estimates of energyrelated carbon dioxide emissions from manufacturing in 2002, based on enduse energy consumption statistics from EIA's Manufacturing Energy Consumption Survey (MECS), which surveys more than 15,000 manufacturing plants every 4 years. The most recent MECS data available are from the 2002 survey. The table on page 24 shows estimates of manufacturing emissions by fuel, based on statistics from the 1991, 1998, and 2002 surveys.

The 1991 MECS reported energy consumption (for fuel and nonfuel purposes) that yielded carbon dioxide emissions from the manufacturing subsector as a whole totaling 1,251.4 million metric tons. The corresponding estimate for 2002 is 1,401.2 million metric tons—an increase of 149.8 million metric tons, representing an average increase of 1.0 percent per year. Over the same interval, the demand for manufacturing products (as measured by gross outputa) increased by 1.3 percent per year. Therefore, the overall carbon intensity of U.S. manufacturing, measured as metric tons of carbon dioxide emitted per million chained 2000 dollars of gross output, was 420.4 in 1991 but had dropped to 358.4 by 2002, representing an average decrease of 1.4 percent per year.

The overall carbon intensity of the U.S. manufacturing subsector is the ratio of its total carbon dioxide emissions (*C*) to manufacturing output (*Y*), as measured by the gross output (in chained 2000 dollars). That ratio (*C/Y*) can be calculated as the product of the subsector's aggregate carbon intensity of energy supply—carbon dioxide emissions (*C*) per unit of energy consumed (*E*)—and its energy intensity—energy consumed (*E*) per unit of gross output (*Y*). That is:

$$C/Y = (C/E) \times (E/Y)$$

For the manufacturing subsector as a whole, energy intensity (the ratio *E/Y*) is a function primarily of the energy intensities of different production groups and their contributions to the total gross output mix in the subsector. The subsector's carbon intensity of energy supply (the ratio *C/E*) is determined primarily by the mix of energy fuel inputs and the mix of fuel and nonfuel (sequestering) uses of the inputs. Thus, the overall carbon intensity of manufacturing (*C/Y*) is a combination of the energy intensity of manufacturing gross output and the carbon intensity of the energy consumed to meet manufacturing energy demand[b] The manufacturing *C/Y* ratio fell by 1.4 percent per year from 1991 to 2002; however, the reduction was largely the result of a structural shift (i.e., a change in relative market shares in the subsector). The energy intensity for the "other manufacturing" category declined by 1.0 percent per year, and at the same time its gross output grew by 3.2 percent per year, from $2.0 trillion in 1991 to $2.9 trillion in 2002 (in chained 2000 dollars), as newer, less energyintensive industries accounted for an increasing share of manufacturing activity. In 1991 the four most energyintensive industries (petroleum, chemicals, primary metals, and paper) accounted for 29.0 percent of total manufacturing gross output, but by 2002 their share had fallen to 23.9 percent. For three of the six manufacturing categories, energy intensity increased from 1991 to 2002 (petroleum by 0.4 percent per year, chemicals 1.5 percent, and nonmetallic minerals 0.1 percent). For paper, primary metals, and other manufacturing, energy intensity declined by 2.0 percent, 0.9 percent, and 1.0 percent per year, respectively.

Carbon Dioxide Emissions from Manufacturing by Industry Group, 2002

Industry Group	NAICS[a] Code	Carbon Dioxide Emissions (Million Metric Tons)	Share of Total Manufacturing Emissions (Percent)	Carbon Intensity of Energy Supply (Million Metric Tons per Quadrillion Btu of Energy Consumed)
Petroleum....	324	304.8	21.8	43.24
Chemicals....	325	311.0	22.2	41.47
Metals.......	331	212.8	15.2	68.72
Paper	322	102.4	7.3	43.35
Minerals	327	91.1	6.5	68.06
Other Manufacturing ...		379.0	27.0	54.58
Total		**1,401.2**	**100.0**	**49.53**

[a]North American Industry Classification System.

Sources: Energy Information Administration, Form EIA846, "Manufacturing Energy Consumption Survey," and Form EIA810, "Monthly Refinery Report" (2002).

Carbon Dioxide Emissions from Manufacturing by Fuel, 1991, 1998, and 2002

Fuel Type	NAICS[a] Code					Other Mfg.	Total
	324	325	331	322	327		
1991							
Carbon Dioxide Emissions (Million Metric Tons)							
Petroleum	149.9	42.1	3.4	13.6	6.4	14.1	**229.6**
Natural Gas	44.2	102.3	37.4	29.1	20.1	73.4	**306.6**
Coal	1.4	25.2	83.2	28.3	27.9	27.2	**193.2**
Electricity	19.8	83.1	94.3	38.0	19.8	192.7	**447.8**
Other	61.0	9.7	3.0	0.2	0.0	0.5	**74.3**
Total	**276.3**	**262.4**	**221.3**	**109.2**	**74.3**	**307.8**	**1,251.4**
Share of Total Gross Output (Percent)	7.1	12.4	4.4	5.1	2.4	68.6	**100.0**
Share of Total Energy Use (Percent)	24.8	23.3	13.2	11.9	4.5	22.3	**100.0**
Share of Total Carbon Dioxide Emissions (Percent)	22.1	21.0	17.7	8.7	5.9	24.6	**100.0**
1998							
Carbon Dioxide Emissions (Million Metric Tons)							
Petroleum	174.8	56.5	3.6	15.1	6.7	13.4	**270.1**
Natural Gas	53.2	127.7	47.9	31.1	23.4	91.5	**374.9**
Coal	0.0	26.9	94.3	25.8	27.7	23.7	**198.3**
Electricity	22.9	103.2	101.8	45.6	24.4	263.7	**561.6**
Other	69.5	4.9	3.4	0.8	0.7	1.6	**80.9**
Total	**320.4**	**319.2**	**251.0**	**118.4**	**82.9**	**393.9**	**1,485.8**
Share of Total Gross Output (Percent)	5.7	11.0	4.2	4.2	2.4	72.5	**100.0**
Share of Total Energy Use (Percent)	25.2	24.0	7.4	10.8	4.1	28.5	**100.0**
Share of Total Carbon Dioxide Emissions (Percent)	21.6	21.5	16.9	8.0	5.6	26.5	**100.0**
2002							
Carbon Dioxide Emissions (Million Metric Tons)							
Petroleum	153.9	70.2	2.4	10.0	11.4	9.6	**257.6**
Natural Gas	46.4	106.2	37.2	26.6	22.3	87.3	**325.9**
Coal	19.3	32.8	72.4	22.5	30.1	25.6	**202.8**
Electricity	24.6	99.4	93.8	42.4	26.8	253.7	**540.7**
Other	60.8	2.4	7.0	0.8	0.4	2.8	**74.2**
Total	**304.8**	**311.0**	**212.8**	**102.4**	**91.1**	**379.0**	**1,401.2**
Share of Total Gross Output (Percent)	5.9	10.5	3.6	3.9	2.2	73.9	**100.0**
Share of Total Energy Use (Percent)	24.9	26.5	10.9	8.4	4.7	24.5	**100.0**
Share of Total Carbon Dioxide Emissions (Percent)	21.8	22.2	15.2	7.3	6.5	27.0	**100.0**

[a]North American Industry Classification System: 324, petroleum products; 325, chemicals; 331, primary metals; 322, paper; 327, nonmetallic minerals (includes stone, clay, and glass).

Notes: Totals may not equal sum of components due to independent rounding. To calculate intensity and consumption values, electricity was calculated as primary electricity: 10,436 Btu per kilowatthour for 1991, 10,197 for 1998, and 10,173 for 2002. These conversion factors represent the average energy input to the generation process for fossilfired utility plants in the United States. See Energy Information Administration, Monthly Energy Review, DOE/EIA0035(2003/09) (Washington, DC, September 2003), Table A6.

Sources: Energy Information Administration, Form EIA846, "Manufacturing Energy Consumption Survey," and Form EIA810, "Monthly Refinery Report" (1991, 1998, and 2002); and U.S. Department of Commerce, Bureau of Economic Analysis, Industry Economics Division.

The mix and quantity of energy fuels consumed by manufacturers (for both fuel and nonfuel uses) affect the subsector's aggregate carbon intensity of energy supply. Overall, manufacturing industries had C/E ratios equal to 50.92 million metric tons carbon dioxide equivalent per quadrillion Btu in 1991 and 49.53 million metric tons carbon dioxide equivalent per quadrillion Btu in 2002; however, the carbon dioxide factors of the various industries differed markedly.

The petroleum and chemical industries both transform some energy products into products that sequester carbon, such as petrochemical feedstocks, asphalt, and plastics. Because of that use, both the petroleum and chemical industries have lower aggregate C/E ratios than the manufacturing average (45.27 and 43.24 million metric tons carbon dioxide equivalent per quadrillion Btu for the petroleum industry and 45.84 and 41.47 for the chemicals industry in 1991 and 2002, respectively).

The paper industry makes extensive use of wood byproducts as an energy source. Carbon dioxide emissions from wood consumption are considered to be zero, because the carbon that is emitted has been sequestered recently, and the regrowing of trees will again sequester an equivalent amount of carbon dioxide. Consequently, the paper industry has a relatively low C/E ratio, at 37.41 and 43.35 million metric tons carbon dioxide equivalent per quadrillion Btu in 1991 and 2002, respectively. In contrast, the primary metals industry, which uses large amounts of coal and other carbonintensive fuels (e.g., electricity), has a high C/E ratio: 68.18 in 1991 and 68.72 in 2002.

Between 1998 and 2002, manufacturing industries had decreases in carbon dioxide emissions associated with their use of electricity (20.9 million metric tons) and natural gas (49.0 million metric tons). Even so, electricity use continues to account for the largest share of manufacturers' energyrelated carbon dioxide emissions: 37.8 percent (561.6 million metric tons) in 1998 and 38.6 percent (540.7 million metric tons) in 2002.

As a result of the above changes in energy intensity, in combination with the structural shift in the subsector, the overall manufacturing energy intensity (E/Y) declined by 1.2 percent per year from 1991 to 2002. When the influence of the structural shift is removed, however, decomposition analysis suggests that the aggregate energy intensity of the manufacturing sector is virtually unchanged[c].

Changes in Key Measures of Carbon Intensity in Manufacturing, 19912002

Industry Group	NAICS[a] Code	1991			2002			Annual Percent Change, 1991-2002		
		E/Y	*C/E*	*C/Y*	*E/Y*	*C/E*	*C/Y*	*E/Y*	*C/E*	*C/Y*
Petroleum	324	29	45.27	1,310.6	30	43.24	1,312.2	0.4	-0.4	0.0
Chemicals	325	15	45.84	708.0	18	41.47	758.0	1.5	-0.9	0.6
Metals	331	25	68.18	1,688.3	22	68.72	1,532.2	-0.9	0.1	-0.9
Paper	322	19	37.41	717.9	15	43.35	668.2	-2.0	1.3	-0.6
Minerals	327	15	67.76	1,048.2	16	68.06	1,058.7	0.1	0.0	0.1
Other Manufacturing		3	56.12	150.8	2	54.58	131.2	-1.0	-0.3	-1.3
Total		**8**	**50.92**	**420.4**	**7**	**49.53**	**358.4**	**-1.2**	**-0.3**	**-1.4**
Total Without Structural Shift		**8**	**NA**	**NA**	**8**	**NA**	**NA**	**-0.1**	**NA**	**NA**

[a]North American Industry Classification System.

Notes: E/Y = energy consumed (thousand Btu) per chained 2000 dollar value of gross output. C/E = million metric tons carbon dioxide emitted per quadrillion Btu of energy consumed. C/Y = metric tons carbon dioxide emitted per million chained 2000 dollars of gross output. NA = not applicable. Annual percent change for E/Y is not statistically significant.

Sources: Energy Information Administration, Form EIA846, "Manufacturing Energy Consumption Survey," and Form EIA810, "Monthly Refinery Report" (1991 and 2002).

Nonfuel Use of Energy Inputs

Nonfuel uses of energy fuels, principally petroleum products, both emit and sequester carbon dioxide over their life cycles. In 2004, nonfuel uses of fossil fuels resulted in emissions of 114.3 MMT carbon dioxide, an increase of 9.1 MMT (8.6 percent) over the 2003 level of 105.2 MMT (Table 12). Carbon dioxide emissions from nonfuel uses, which represent about 2 percent of total U.S. carbon dioxide emissions, have grown by an average of 1.1 percent annually from their 1990 level of 98.1 MMT. Emissions from nonfuel uses of petroleum products in 2004 were 89.0 MMT in the industrial sector and 5.6 MMT in the transportation sector. Within the industrial petroleum products category, the leading carbon dioxide emission sources were petrochemical feedstocks at 41.4 MMT and LPG at 19.3 MMT. Nonfuel uses of natural gas resulted in emissions of 19.2 MMT carbon dioxide in 2004.

In 2004, carbon sequestration through nonfuel uses of fossil fuels totaled 316.6 $MMTCO_2e$ (Table 13). The vast majority was sequestered in petroleumbased products, including 287.7 $MMTCO_2e$ in the industrial sector and 5.6 $MMTCO_2e$ in the transportation sector sequestered through the use of petroleumbased lubricants. Smaller amounts of carbon were sequestered in naturalgasbased products (21.8 $MMTCO_2e$) and coalbased products (1.5 $MMTCO_2e$). The main products that sequester carbon include asphalt and road oil (98.6 $MMTCO_2e$), LPG (77.7 $MMTCO_2e$), and feedstocks for plastics and other petrochemicals (69.2 $MMTCO_2e$). The amount sequestered in 2004 was 7.7 percent more than in 2003, when 294.0 $MMTCO_2e$ was sequestered. Since 1990, the annual sequestration of carbon in this manner has increased by 65.4 $MMTCO_2e$ or 26 percent. This translates to an average annual growth rate of 1.7 percent.

Adjustments to Energy Consumption

Total energy consumption and the carbon dioxide emissions upon which they are based correspond to EIA's coverage of energy consumption, which includes the 50 States and the District of Columbia. Under the United Nations Framework Convention on Climate Change (UNFCCC), however, the United States is also responsible for counting emissions emanating from its Territories, and their emissions are added to the U.S. total. Conversely, because the Intergovernmental Panel on Climate Change (IPCC) definition of energy consumption excludes international bunker fuels from the statistics of all countries, emissions from international bunker fuels are subtracted from the U.S. total. Additionally, military bunker fuels are subtracted because they are also excluded by the IPCC from the national total. These sources and subtractions are enumerated and described as "adjustments to energy."

U.S. Territories

Energyrelated carbon dioxide emissions for the U.S. Territories are added as an adjustment in keeping with IPCC guidelines for national emissions inventories. The Territories included are Puerto Rico, the U.S. Virgin Islands, American Samoa, Guam, the U.S. Pacific Islands, and Wake Island. Most of these emissions are from petroleum products; however, Puerto Rico and the Virgin Islands consume coal in addition to petroleum products. For 2004, total energyrelated carbon dioxide emissions from the U.S. Territories are estimated at 61.7 MMT (Table 5).

International Bunker Fuels

In keeping with the IPCC guidelines for estimating national greenhouse gas emissions, carbon dioxide emissions from international bunker fuels are subtracted from the estimate of total U.S. energyrelated emissions of carbon dioxide. Purchases of distillate and residual fuels by foreignbound ships at U.S. seaports, as well as jet fuel purchases by international air carriers at U.S. airports, form the basis of the estimate for bunker fuels. Additionally, U.S. military operations for which fuel was originally purchased in the United States but consumed in international waters or airspace are subtracted from the total, because they are also considered international bunker fuels under this definition.

For 2003, the carbon dioxide emissions estimate for military bunker fuels was 9.2 MMT.[59] In 2004, approximately 93.6 MMT carbon dioxide was emitted in total from international bunker fuels, including 84.4 MMT attributed to civilian consumption of bunker fuels. The total amount is subtracted from the U.S. total in Table 5. Just over onehalf of the carbon dioxide emissions associated with international bunker fuels comes from the combustion of jet fuels; residual and distillate fuels account for the other half, with most coming from residual fuel.

Other Carbon Dioxide Emissions

U.S. Carbon Dioxide Emissions from Other Sources, **1990-2004**

Estimated 2004 Emissions (Million Metric Tons Carbon Dioxide)	105.0
Change Compared to 2003 (Million Metric Tons Carbon Dioxide)	2.8
Change from 2003 *(Percent)*	*2.8%*
Change Compared to 1990 (Million Metric Tons Carbon Dioxide)	16.7
Change from 1990 *(Percent)*	*18.9%*

Energy Production

In addition to emissions resulting from fossil energy consumed, oil and gas production leads to emissions of carbon dioxide from sources other than the combustion of those marketed fossil fuels. The two energy production sources estimated for this report are:

- Flared natural gas (gas burned at the production site), which is flared either because the cost of bringing the gas to market is prohibitive or because the gas is of insufficient quality to sell.
- Carbon dioxide scrubbed from natural gas to improve its heat content and quality and subsequently vented to the atmosphere.

Because many States require flaring of natural gas, EIA assumes that all gas reported under the category "Vented and Flared" is actually flared and therefore should be counted as carbon dioxide emissions rather than methane emissions. In 2004, about 5.9 MMT carbon dioxide was emitted in this way (Table 5).

By computing the difference between the estimated carbon dioxide content of raw gas and the carbon dioxide content of pipeline gas, the amount of carbon dioxide that has been removed (scrubbed) in order to improve the heat content and quality of natural gas can be calculated. This amount was about 17.8 MMT in 2004 (Table 5).

Information on energy production sources that are excluded from this report because of insufficient data are available in Energy Information Administration, *Documentation for Emissions of Greenhouse Gases in the United States 2003*.[60]

Industrial Process Emissions

Industrial emissions of carbon dioxide not caused by the combustion of fossil fuels accounted for only 1.2 percent (73.5 MMT) of total U.S. carbon dioxide emissions in 2004 (Table 14). Processrelated emissions from industrial sources depend largely on the level of activity in the construction industries and on production at oil and gas wells. These sources include limestone and dolomite calcination, soda ash manufacture and consumption, carbon dioxide manufacture, cement manufacture, and aluminum production.

Estimated industrial process emissions of carbon dioxide in 2004 totaled 73.5 MMT, 13.4 MMT (22 percent) higher than in 1990 and 2.7 MMT (3.9 percent) higher than in 2003 (Table 14). Of the total estimate for carbon dioxide emissions from industrial processes in 2004, 61 percent is attributed to cement manufacture. When calcium carbonate is heated (calcined) in a kiln, it is converted to lime and carbon dioxide. The lime is combined with other materials to produce clinker (an intermediate product from which cement is made), and the carbon dioxide is released to the atmosphere. In 2004, the United States produced an estimated 95.0 million metric tons of cement,[61] resulting in the direct release of 44.8 MMT into the atmosphere. This calculation is independent of the carbon dioxide released by the combustion of energy fuels consumed in making cement. The estimate for 2004 represents an increase in carbon dioxide emissions of 11.5 MMT (34 percent) compared with 1990 and an increase of about 1.6 MMT (3.8 percent) compared with 2003.

Collectively, in 2004, industrial processes other than cement manufacture emitted 28.7 MMT carbon dioxide. Limestone manufacture and consumption emitted 19.1 MMT, soda ash manufacture 3.8 MMT, aluminum manufacture 3.7 MMT, carbon dioxide manufacture 1.5 MMT, and soda ash consumption 0.6 MMT.

Waste Combustion

Waste that is combusted contains, on average, a portion that is composed of plastics, synthetic rubber, synthetic fibers, and carbon black. The carbon in these plastics has normally been accounted for as sequestered carbon, as reported in Table 13; however, according to the IPCC, emissions from the plastics contained in municipal solid waste must be counted in total national emissions inventories. The U.S. Environmental Protection Agency (EPA) estimates that plastics and other nonbiogenic materials in combusted waste produced emissions of about 18.8 MMT carbon dioxide in 2003 (about 11 MMT from gridconnected power generation).[62] The EPA s 2003 value is used in this report as an estimate for 2004. The difference between the estimated total and EIA s estimate for the electric power sector is reported in Table 5. For 2004, the difference is 7.8 MMT carbon dioxide.

Table 5. U.S. Carbon Dioxide Emissions from Energy and Industry, 1990 and 1996-2004 (Million Metric Tons Carbon Dioxide)

Fuel Type or Process	1990	1996	1997	1998	1999	2000	2001	2002	2003	P2004
Energy Consumption										
Petroleum	2,179.6	2,278.6	2,296.6	2,348.7	2,416.3	2,438.5	2,475.6	2,457.5	2,497.5	2,595.2
Coal	1,783.5	1,960.8	2,005.5	2,023.0	2,024.2	2,115.7	2,055.6	2,049.8	2,090.2	2,090.0
Natural Gas	1,027.2	1,206.8	1,212.3	1,189.3	1,191.9	1,237.5	1,186.0	1,225.7	1,196.1	1,203.4
Renewables[a]	6.2	10.3	10.9	10.6	10.8	10.5	11.2	13.0	11.7	11.4
Energy Subtotal	**4,996.6**	**5,456.5**	**5,525.2**	**5,571.6**	**5,643.2**	**5,802.3**	**5,728.4**	**5,746.0**	**5,795.5**	**5,899.9**
Nonfuel Use Emissions[b]	**98.1**	**105.6**	**110.5**	**118.6**	**124.1**	**110.0**	**104.9**	**107.0**	**105.2**	**114.3**
Nonfuel Use Sequestration[c]	**251.2**	**291.2**	**303.4**	**314.7**	**325.8**	**308.1**	**293.9**	**298.3**	**294.0**	**316.6**
Adjustments to Energy										
U.S. Territories (+)	31.1	37.7	39.0	41.3	40.7	42.2	53.6	52.2	58.4	61.7
Military Bunker Fuels (-)	13.6	8.9	9.6	10.0	9.8	7.8	8.2	8.1	9.2	9.2
International Bunker Fuels (-)	100.1	93.3	100.1	104.9	97.4	93.5	89.6	81.2	75.0	84.4
Total Energy Adjustments	**-82.6**	**-64.5**	**-70.7**	**-73.6**	**-66.5**	**-59.1**	**-44.2**	**-37.1**	**-25.9**	**-31.9**
Adjusted Energy Subtotal	**4,914.0**	**5,392.0**	**5,454.6**	**5,498.0**	**5,576.8**	**5,743.1**	**5,684.2**	**5,709.0**	**5,769.6**	**5,868.0**
Other Sources										
Natural Gas Flaring	9.1	16.5	15.5	6.2	6.7	5.5	5.9	6.0	5.9	5.9
Carbon Dioxide in Natural Gas	14.0	17.8	18.0	18.0	17.8	18.2	18.6	17.9	18.0	17.8
Cement Production	33.3	37.2	38.4	39.3	40.1	41.3	41.5	43.0	43.2	44.8
Other Industrial	26.8	29.0	29.2	29.7	29.3	29.4	27.4	26.4	27.6	28.7
Waste Combustion	5.1	7.3	7.3	6.9	7.2	7.9	8.0	6.2	7.5	7.8
Total Other Sources	**88.3**	**107.7**	**108.5**	**100.1**	**101.2**	**102.3**	**101.3**	**99.5**	**102.2**	**105.0**
Total	**5,002.3**	**5,499.7**	**5,563.0**	**5,598.1**	**5,677.9**	**5,845.5**	**5,785.5**	**5,808.5**	**5,871.8**	**5,973.0**

[a]Includes emissions from electricity generation using municipal solid waste and geothermal energy.

[b]Emissions from nonfuel use are included in the fuel categories above. See Table 12 for details by fuel category.

[c]See Table 13 for details by fuel.

P = preliminary data.

Notes: Data in this table are revised from the data contained in the previous EIA report, *Emissions of Greenhouse Gases in the United States 2003,* DOE/EIA0573(2003) (Washington, DC, December 2004). Totals may not equal sum of components due to independent rounding. Adjusted energy total includes U.S. Territories.

Sources: EIA estimates presented in this chapter.

Table 6. U.S. Carbon Dioxide Emissions from Energy Consumption by EndUse Sector, 1990 and 1996-2004 (Million Metric Tons Carbon Dioxide)

End-Use Sector	1990	1996	1997	1998	1999	2000	2001	2002	2003	P2004
Residential	953.7	1,090.2	1,082.0	1,088.5	1,110.9	1,171.8	1,168.3	1,190.4	1,213.2	1,212.0
Commercial	780.7	872.1	916.1	934.1	947.3	1,005.9	1,026.2	1,019.8	1,021.1	1,024.2
Industrial	1,692.2	1,788.8	1,804.4	1,791.2	1,779.1	1,780.3	1,698.4	1,671.0	1,685.6	1,730.2
Transportation	1,569.9	1,705.3	1,722.7	1,757.9	1,806.0	1,844.2	1,835.5	1,864.9	1,875.7	1,933.7
Total[a]	**4,996.6**	**5,456.5**	**5,525.2**	**5,571.6**	**5,643.2**	**5,802.3**	**5,728.4**	**5,746.0**	**5,795.5**	**5,899.9**
Electric Power	1,803.1	2,007.1	2,076.3	2,164.9	2,175.8	2,279.3	2,248.2	2,250.6	2,278.8	2,298.6

P = preliminary data.

[a]Includes emissions from nonfuel uses of fossil fuels. See Table 12 for details by fuel category.

Notes: Data in this table are revised from the data contained in the previous EIA report, *Emissions of Greenhouse Gases in the United States 2003,* DOE/EIA0573(2003) (Washington, DC, December 2004). Totals may not equal sum of components due to independent rounding. Electric power sector emissions are distributed across the enduse sectors. Emissions allocated to sectors are unadjusted. Adjustments are made to total emissions only (Table 5).

Sources: EIA estimates presented in this chapter.

Table 7. U.S. Carbon Dioxide Emissions from Residential Sector Energy Consumption, 1990 and 1996-2004 (Million Metric Tons Carbon Dioxide)

Fuel	1990	1996	1997	1998	1999	2000	2001	2002	2003	P2004
Petroleum										
Liquefied Petroleum Gas . .	22.6	29.3	28.6	26.9	33.1	34.9	33.2	33.6	32.4	33.3
Distillate Fuel	70.8	67.1	63.3	55.9	59.9	65.5	65.8	62.3	65.5	68.4
Kerosene	4.6	6.4	6.6	7.8	8.0	6.8	6.8	4.3	5.4	6.4
Petroleum Subtotal	**98.0**	**102.7**	**98.5**	**90.5**	**101.0**	**107.2**	**105.8**	**100.2**	**103.3**	**108.0**
Coal	2.9	1.5	1.5	1.2	1.3	1.0	1.1	1.0	0.9	1.0
Natural Gas	238.6	284.6	270.6	246.6	256.4	269.2	259.0	265.6	276.9	265.5
Electricity[a]	614.2	701.3	711.4	750.2	752.2	794.4	802.5	823.5	832.0	837.3
Total.	**953.7**	**1,090.2**	**1,082.0**	**1,088.5**	**1,110.9**	**1,171.8**	**1,168.3**	**1,190.4**	**1,213.2**	**1,212.0**

[a]Share of total electric power sector carbon dioxide emissions weighted by sales to the residential sector.

P = preliminary data.

Notes: Data in this table are revised from the data contained in the previous EIA report, *Emissions of Greenhouse Gases in the United States 2003,* DOE/EIA0573(2003) (Washington, DC, December 2004). Totals may not equal sum of components due to independent rounding.

Sources: EIA estimates presented in this chapter.

Table 8. U.S. Carbon Dioxide Emissions from Commercial Sector Energy Consumption, 1990 and 1996-2004 (Million Metric Tons Carbon Dioxide)

Fuel	1990	1996	1997	1998	1999	2000	2001	2002	2003	P2004
Petroleum										
Motor Gasoline	7.8	1.9	3.0	2.7	2.0	3.1	2.6	3.2	3.2	3.3
Liquefied Petroleum Gas. . .	4.0	5.2	5.0	4.7	5.8	6.2	5.9	5.9	5.7	5.9
Distillate Fuel.	38.8	35.0	32.1	31.1	31.8	35.6	36.8	32.2	34.9	36.4
Residual Fuel	17.9	10.7	8.7	6.6	5.7	6.6	6.5	6.2	8.7	10.6
Kerosene.	0.8	1.5	1.8	2.2	1.9	2.1	2.2	1.1	1.4	1.7
Petroleum Subtotal.	**69.4**	**54.2**	**50.6**	**47.4**	**47.2**	**53.6**	**54.0**	**48.7**	**53.9**	**57.9**
Coal.	11.6	11.3	12.2	9.5	9.6	8.1	8.6	8.6	7.9	8.2
Natural Gas.	142.4	171.6	174.5	163.6	165.3	171.8	164.9	170.8	175.4	162.7
Electricity[a].	557.2	635.0	678.8	713.6	725.1	772.4	798.6	791.8	783.8	795.4
Total	**780.7**	**872.1**	**916.1**	**934.1**	**947.3**	**1,005.9**	**1,026.2**	**1,019.8**	**1,021.1**	**1,024.2**

[a]Share of total electric power sector carbon dioxide emissions weighted by sales to the commercial sector.

P = preliminary data.

Notes: Data in this table are revised from the data contained in the previous EIA report, *Emissions of Greenhouse Gases in the United States 2003,* DOE/EIA0573(2003) (Washington, DC, December 2004). Totals may not equal sum of components due to independent rounding.

Sources: EIA estimates presented in this chapter.

Table 9. U.S. Carbon Dioxide Emissions from Industrial Sector Energy Consumption, 1990 and 1996-2004 (Million Metric Tons Carbon Dioxide)

Fuel	1990	1996	1997	1998	1999	2000	2001	2002	2003	P2004
Petroleum										
Motor Gasoline	13.0	14.0	14.9	14.0	10.6	10.5	20.7	21.7	21.9	22.4
Liquefied Petroleum Gas	39.9	47.4	49.3	40.3	49.3	57.9	50.1	55.8	54.6	56.6
Distillate Fuel	83.1	85.6	86.7	87.3	85.5	86.5	93.7	86.8	81.9	85.4
Residual Fuel	30.3	24.1	20.5	15.7	14.0	15.4	16.6	12.6	15.0	18.9
Asphalt and Road Oil	0.0	0.0	0.0	0.0	0.0	0.0	0.0	0.0	0.0	0.0
Lubricants	6.8	6.3	6.7	7.0	7.1	7.0	6.4	6.3	5.8	5.9
Kerosene	0.9	1.3	1.3	1.6	0.9	1.1	1.7	1.0	1.2	1.5
Petroleum Coke	81.2	84.0	79.5	97.6	109.0	87.6	97.2	94.8	91.6	110.0
Other Petroleum	125.1	130.5	135.9	123.6	128.0	115.5	130.0	127.3	137.3	139.8
Petroleum Subtotal	**380.3**	**393.2**	**394.9**	**387.1**	**404.4**	**381.6**	**416.4**	**406.3**	**409.4**	**440.6**
Coal	**249.5**	**219.0**	**215.3**	**203.8**	**199.2**	**202.9**	**195.9**	**180.3**	**183.2**	**181.0**
Coal Coke Net Imports	**0.5**	**3.2**	**5.3**	**7.4**	**6.5**	**7.2**	**4.0**	**5.8**	**5.8**	**5.8**
Natural Gas	**433.5**	**505.7**	**506.1**	**495.0**	**473.6**	**479.7**	**438.7**	**446.6**	**428.8**	**441.9**
Electricity[a]	**628.5**	**667.6**	**682.9**	**697.8**	**695.3**	**709.0**	**643.4**	**631.9**	**658.5**	**660.9**
Total[b]	**1,692.2**	**1,788.8**	**1,804.4**	**1,791.2**	**1,779.1**	**1,780.3**	**1,698.4**	**1,671.0**	**1,685.6**	**1,730.2**

[a]Share of total electric power sector carbon dioxide emissions weighted by sales to the industrial sector.
[b]Includes emissions from nonfuel uses of fossil fuels. See Table 12 for details by fuel category.
P = preliminary data.

Notes: Data in this table are revised from the data contained in the previous EIA report, *Emissions of Greenhouse Gases in the United States 2003,* DOE/EIA0573(2003) (Washington, DC, December 2004).Totals may not equal sum of components due to independent rounding.

Sources: EIA estimates presented in this chapter.

Table 10. U.S. Carbon Dioxide Emissions from Transportation Sector Energy Consumption, 1990 and 1996-2004 (Million Metric Tons Carbon Dioxide)

Fuel	1990	1996	1997	1998	1999	2000	2001	2002	2003	P2004
Petroleum										
Motor Gasoline	955.2	1,034.0	1,042.5	1,072.9	1,099.9	1,106.0	1,110.9	1,137.6	1,141.4	1,162.6
Liquefied Petroleum Gas	1.3	0.9	0.8	1.0	0.8	0.7	0.8	0.8	0.8	0.8
Jet Fuel	220.4	229.8	232.1	235.6	242.9	251.2	240.4	234.4	229.1	237.4
Distillate Fuel	265.1	323.7	338.4	348.4	362.2	374.0	383.2	390.5	410.3	428.2
Residual Fuel	79.3	66.4	55.5	52.6	51.9	64.2	54.2	52.8	44.5	54.6
Lubricants[a]	6.5	6.0	6.3	6.6	6.7	6.6	6.0	6.0	5.5	5.6
Aviation Gasoline	3.1	2.6	2.7	2.4	2.7	2.5	2.4	2.3	2.1	2.1
Petroleum Subtotal	**1,530.9**	**1,663.3**	**1,678.4**	**1,719.5**	**1,767.1**	**1,805.2**	**1,797.9**	**1,824.5**	**1,833.8**	**1,891.3**
Coal	**0.0**	**0.0**	**0.0**	**0.0**	**0.0**	**0.0**	**0.0**	**0.0**	**0.0**	**0.0**
Natural Gas	**35.9**	**38.9**	**41.1**	**35.1**	**35.6**	**35.5**	**33.9**	**37.1**	**37.3**	**37.4**
Electricity[b]	**3.2**	**3.2**	**3.2**	**3.3**	**3.2**	**3.6**	**3.7**	**3.4**	**4.6**	**5.0**
Total	**1,569.9**	**1,705.3**	**1,722.7**	**1,757.9**	**1,806.0**	**1,844.2**	**1,835.5**	**1,864.9**	**1,875.7**	**1,933.7**

[a]Includes emissions from nonfuel uses of fossil fuels. See Table 12 for details by fuel category.
[b]Share of total electric power sector carbon dioxide emissions weighted by sales to the transportation sector.
P = preliminary data.

Notes: Data in this table are revised from the data contained in the previous EIA report, *Emissions of Greenhouse Gases in the United States 2003,* DOE/EIA0573(2003) (Washington, DC, December 2004). Totals may not equal sum of components due to independent rounding.

Sources: EIA estimates presented in this chapter.

Table 11. U.S. Carbon Dioxide Emissions from Electric Power Sector Energy Consumption, 1990 and 1996-2004 (Million Metric Tons Carbon Dioxide)

Fuel	1990	1996	1997	1998	1999	2000	2001	2002	2003	P2004
Petroleum										
Heavy Fuel Oil	91.0	49.5	56.4	82.7	75.7	68.8	79.3	51.7	68.3	69.1
Light Fuel Oil..........	7.0	7.9	8.0	9.8	10.1	12.6	12.3	9.2	11.6	8.0
Petroleum Coke	2.9	7.6	9.7	11.7	10.7	9.5	9.9	16.4	16.4	19.4
Petroleum Subtotal[a]..	**100.9**	**65.1**	**74.2**	**104.1**	**96.5**	**90.9**	**101.5**	**77.9**	**97.1**	**97.4**
Coal	**1,519.1**	**1,725.8**	**1,771.3**	**1,801.1**	**1,807.6**	**1,896.6**	**1,846.0**	**1,854.1**	**1,892.4**	**1,893.9**
Natural Gas	**176.9**	**206.0**	**220.0**	**249.1**	**260.9**	**281.4**	**289.5**	**305.6**	**277.6**	**295.9**
Municipal Solid Waste ..	**5.8**	**9.9**	**10.5**	**10.2**	**10.4**	**10.1**	**10.8**	**12.6**	**11.3**	**11.0**
Geothermal	**0.4**	**0.4**	**0.4**	**0.4**	**0.4**	**0.4**	**0.4**	**0.4**	**0.4**	**0.4**
Total................	**1,803.1**	**2,007.1**	**2,076.3**	**2,164.9**	**2,175.8**	**2,279.3**	**2,248.2**	**2,250.6**	**2,278.8**	**2,298.6**

P = preliminary data.

[a]Includes small amounts of other petroleum liquids, such as jet fuel and waste oil.

Notes: Data in this table are revised from the data contained in the previous EIA report, *Emissions of Greenhouse Gases in the United States 2003,* DOE/EIA0573(2003) (Washington, DC, December 2004). Emissions for total fuel consumption are allocated to enduse sectors in proportion to electricity sales. Totals may not equal sum of components due to independent rounding.

Sources: EIA estimates presented in this chapter.

Table 12. U.S. Carbon Dioxide Emissions from Nonfuel Use of Energy Fuels, 1990 and 1996-2004 (Million Metric Tons Carbon Dioxide)

End Use and Type	1990	1996	1997	1998	1999	2000	2001	2002	2003	P2004
Industrial										
Petroleum										
Liquefied Petroleum Gases...	14.8	20.3	20.6	21.5	22.4	20.4	19.1	19.9	19.0	19.3
Distillate Fuel	0.3	0.3	0.4	0.4	0.4	0.4	0.4	0.4	0.4	0.4
Residual Fuel	1.8	2.1	2.1	2.2	2.2	2.2	2.2	2.2	2.2	2.2
Asphalt and Road Oil	0.0	0.0	0.0	0.0	0.0	0.0	0.0	0.0	0.0	0.0
Lubricants	6.8	6.3	6.7	7.0	7.1	7.0	6.4	6.3	5.8	5.9
Other *(Subtotal)*	*51.2*	*51.4*	*55.2*	*60.6*	*66.2*	*53.7*	*51.4*	*52.7*	*52.5*	*61.1*
Pentanes Plus...........	1.1	4.2	4.0	2.7	3.5	3.1	2.7	2.3	2.2	2.3
Petrochemical Feed	33.3	34.3	40.0	39.3	37.8	36.4	32.6	33.2	36.2	41.4
Petroleum Coke	9.0	7.5	5.9	10.8	14.4	7.1	10.5	9.7	8.3	13.8
Special Naphtha	7.8	5.4	5.3	7.8	10.6	7.1	5.7	7.5	5.9	3.7
Waxes and Miscellaneous ..	0.0	0.0	0.0	0.0	0.0	0.0	0.0	0.0	0.0	0.0
Petroleum Subtotal	*74.9*	*80.5*	*85.0*	*91.7*	*98.3*	*83.8*	*79.5*	*81.5*	*80.0*	*89.0*
Coal.....................	0.5	0.7	0.6	0.6	0.6	0.6	0.6	0.5	0.5	0.5
Natural Gas...............	16.2	18.5	18.5	19.7	18.5	19.0	18.8	19.1	19.2	19.2
Industrial Subtotal	***91.6***	***99.6***	***104.1***	***112.0***	***117.4***	***103.4***	***98.9***	***101.1***	***99.7***	***108.7***
Transportation										
Lubricants	6.5	6.0	6.3	6.6	6.7	6.6	6.0	6.0	5.5	5.6
Total	**98.1**	**105.6**	**110.5**	**118.6**	**124.1**	**110.0**	**104.9**	**107.0**	**105.2**	**114.3**

P = preliminary data.

Notes: Emissions from nonfuel use of energy fuels are included in the energy consumption tables in this chapter. Data in this table are revised from unpublished data used to produce the previous EIA report, *Emissions of Greenhouse Gases in the United States 2003,* DOE/EIA0573(2003) (Washington, DC, December 2004). Totals may not equal sum of components due to independent rounding.

Sources: EIA estimates presented in this chapter.

Table 13. U.S. Carbon Sequestered by Nonfuel Use of Energy Fuels, 1990 and 1996-2004 (Million Metric Tons Carbon Dioxide Equivalent)

End Use and Type	1990	1996	1997	1998	1999	2000	2001	2002	2003	P2004
Industrial										
Petroleum										
Liquefied Petroleum Gases	59.3	81.7	82.7	86.4	89.9	82.1	76.7	79.9	76.3	77.7
Distillate Fuel	0.3	0.3	0.4	0.4	0.4	0.4	0.4	0.4	0.4	0.4
Residual Fuel	1.9	2.1	2.2	2.2	2.2	2.2	2.2	2.2	2.2	2.2
Asphalt and Road Oil	88.5	88.9	92.5	95.5	100.1	96.4	95.0	93.7	92.2	98.6
Lubricants	6.9	6.4	6.8	7.1	7.2	7.0	6.5	6.4	5.9	6.0
Other *(Subtotal)*	*72.0*	*85.2*	*90.7*	*92.7*	*94.9*	*88.7*	*83.9*	*86.4*	*88.1*	*102.8*
Pentanes Plus	4.4	16.9	16.0	10.9	14.0	12.7	10.8	9.2	9.0	9.1
Petrochemical Feed	46.0	50.6	58.4	59.1	55.5	57.7	50.7	55.1	59.2	69.2
Petroleum Coke	9.1	7.6	6.0	10.9	14.5	7.2	10.6	9.8	8.3	13.9
Special Naphtha	0.0	0.0	0.0	0.0	0.0	0.0	0.0	0.0	0.0	0.0
Waxes and Miscellaneous	12.5	10.1	10.4	11.8	10.9	11.2	11.8	12.2	11.5	10.6
Petroleum Subtotal	*228.8*	*264.6*	*275.2*	*284.3*	*294.8*	*277.0*	*264.7*	*269.1*	*265.2*	*287.7*
Coal	1.4	2.0	1.9	1.8	1.8	1.8	1.7	1.5	1.5	1.5
Natural Gas	14.4	18.5	19.8	21.9	22.5	22.7	21.3	21.7	21.8	21.8
Industrial Subtotal	***244.7***	***285.2***	***297.0***	***308.0***	***319.1***	***301.5***	***287.8***	***292.3***	***288.5***	***311.0***
Transportation										
Lubricants	6.5	6.0	6.4	6.7	6.8	6.7	6.1	6.0	5.6	5.6
Total	**251.2**	**291.2**	**303.4**	**314.7**	**325.8**	**308.1**	**293.9**	**298.3**	**294.0**	**316.6**

P = preliminary data.

Notes: Emissions from nonfuel use of energy fuels are included in the energy consumption tables in this chapter. Data in this table are revised from the data contained in the previous EIA report, *Emissions of Greenhouse Gases in the United States 2003,* DOE/EIA0573(2003) (Washington, DC, December 2004). Totals may not equal sum of components due to independent rounding.

Sources: EIA estimates presented in this chapter.

Table 14. U.S. Carbon Dioxide Emissions from Industrial Processes, 1990 and 1996-2004 (Million Metric Tons Carbon Dioxide)

Source	1990	1996	1997	1998	1999	2000	2001	2002	2003	P2004
Cement Manufacture										
Clinker Production	32.6	36.3	37.6	38.4	39.2	40.4	40.5	42.0	42.2	43.8
Masonry Cement	0.1	0.1	0.1	0.1	0.1	0.1	0.1	0.1	0.1	0.1
Cement Kiln Dust	0.7	0.7	0.8	0.8	0.8	0.8	0.8	0.8	0.8	0.9
Cement Subtotal	**33.3**	**37.2**	**38.4**	**39.3**	**40.1**	**41.3**	**41.5**	**43.0**	**43.2**	**44.8**
Other Industrial										
Limestone Consumption										
Lime Manufacture	12.4	15.1	15.5	15.8	15.5	15.4	14.8	14.1	15.1	16.0
Iron Smelting	1.7	1.1	1.1	1.1	1.0	1.1	1.0	0.9	0.9	0.9
Steelmaking	0.3	0.4	0.3	0.4	0.3	0.5	0.6	0.5	0.4	0.4
Copper Refining	0.1	0.2	0.2	0.2	0.2	0.1	0.1	0.1	0.1	0.1
Glass Manufacture	0.1	0.2	0.1	0.0	0.2	0.2	0.1	0.1	0.2	0.2
Flue Gas Desulfurization	0.7	0.9	1.0	1.0	1.1	1.2	1.4	1.4	1.3	1.4
Dolomite Manufacture	0.5	0.3	0.0	0.3	0.1	0.1	0.1	0.1	0.1	0.1
Limestone Subtotal	***15.9***	***18.3***	***18.3***	***18.7***	***18.3***	***18.6***	***18.1***	***17.1***	***18.0***	***19.1***
Soda Ash Manufacture	3.4	3.8	3.9	3.8	3.7	3.6	3.6	3.5	3.6	3.8
Soda Ash Consumption										
Glass Manufacture	0.1	0.2	0.1	0.0	0.2	0.2	0.1	0.1	0.2	0.2
Flue Gas Desulfurization	0.1	0.1	0.1	0.1	0.1	0.1	0.2	0.1	0.1	0.1
Sodium Silicate	0.2	0.2	0.3	0.3	0.2	0.2	0.3	0.2	0.3	0.3
Sodium Tripolyphosphate	0.1	0.1	0.1	0.1	0.1	0.1	0.1	0.1	0.1	0.1
Soda Ash Subtotal	***0.5***	***0.6***	***0.6***	***0.5***	***0.6***	***0.6***	***0.5***	***0.4***	***0.6***	***0.6***
Carbon Dioxide Manufacture	0.9	1.1	1.1	1.2	1.2	1.3	1.3	1.4	1.5	1.5
Aluminum Manufacture	5.9	5.2	5.3	5.4	5.5	5.4	3.9	4.0	4.0	3.7
Shale Oil Production	0.2	*	*	*	*	*	*	*	*	*
Other Industrial Subtotal	**26.8**	**29.0**	**29.2**	**29.7**	**29.3**	**29.4**	**27.4**	**26.4**	**27.6**	**28.7**
Total	**60.1**	**66.1**	**67.6**	**69.0**	**69.4**	**70.7**	**68.9**	**69.4**	**70.8**	**73.5**

*Less than 50,000 metric tons carbon equivalent.

P = preliminary data.

Notes: Data in this table are revised from the data contained in the previous EIA report, Emissions of Greenhouse Gases in the United States 2003, DOE/EIA0573(2003) (Washington, DC, December 2004). Totals may not equal sum of components due to independent rounding.

Sources: EIA estimates presented in this chapter.

METHANE EMISSIONS

Overview

U.S. Anthropogenic Methane Emissions, 1990-2004

	Methane	Carbon Dioxide Equivalent
Estimated 2004 Emissions (Million Metric Tons)	27.8	639.5
Change Compared to 2003 (Million Metric Tons)	0.2	5.6
Change from 2003 *(Percent)*	*0.9%*	*0.9%*
Change Compared to 1990 (Million Metric Tons)	-3.6	-81.9
Change from 1990 *(Percent)*	*-11.4%*	*-11.4%*

U.S. anthropogenic methane emissions in 2004 totaled 639.5 million metric tons carbon dioxide equivalent ($MMTCO_2e$),[63] or 27.8 million metric tons of methane, representing 9.0 percent of total U.S. greenhouse gas emissions. U.S. methane emissions in 2004 were 0.9 percent (5.6 $MMTCO_2e$) higher than their 2003 level of 633.9 $MMTCO_2e$ (Table 15), primarily as a result of an increase in emissions from landfills and smaller increases in emissions associated with animal waste, rice cultivation, and coal mining.

Principal Sources of U.S. Anthropogenic Methane Emissions, 1990-2004

Source	Million Metric Tons CO_2e 1990	2004	*Percent Change* 1990-2004	2003-2004
Energy	275.04	256.31	*-6.8%*	*
Waste Management	270.21	198.19	*-26.7%*	*2.4%*
Agriculture	173.42	182.25	*5.1%*	*0.5%*
Industrial Processes	2.70	2.70	*0.2%*	*4.1%*

*Less than 0.05 percent.

U.S. emissions of methane in 2004 were 11.4 percent (81.9 $MMTCO_2e$) below their 1990 level of 721.4 $MMTCO_2e$ (Figure 2). In addition to a reduction of 74.4 $MMTCO_2e$ (29 percent) in methane emissions from landfills since 1990, there was also a decrease of 29.5 $MMTCO_2e$ (30 percent) in methane emissions from coal mines over the same period (Table

16). The 30 percent decline in emissions from coal mining was the result of a 150percent increase in methane recovery from coal mines and a shift in production away from gassy mines.

Methane emission estimates are much more uncertain than carbon dioxide emission estimates. Methane emissions usually are accidental or incidental to biological processes and may not be metered in any systematic way.[64] Thus, methane emission estimates must often rely on proxy measurements.

Estimated U.S. anthropogenic methane emissions for 2004 are based on incomplete data for several key sources; thus, the overall estimate is likely to be revised. Because emissions from three of these sources—coal mining, natural gas systems, and landfills—represented more than threefifths of all U.S. methane emissions, comparisons between 2003 and 2004 numbers are more likely to be valid in terms of their direction than their magnitude. For example, because 2004 data on waste generation are not yet available, waste generation has been estimated from a simple regression equation with economic output as the independent variable. Less critical but still important data are also unavailable for natural gas systems, such as miles of gas transmission and distribution pipeline.

Energy Sources

U.S. methane emissions from energy sources in 2004 are estimated at 256.3 $MMTCO_2e$, equivalent to 40 percent of total methane emissions. The 2004 emission level is nearly unchanged from the 2003 level of 256.4 $MMTCO_2e$. Total methane emissions from energy sources in 2004 were 18.7 $MMTCO_2e$ below their 1990 level of 275.0 $MMTCO_2e$.

The drop in methane emissions from energy sources since 1990 can be traced primarily to an overall reduction in emissions from coal mines and secondarily to lower emissions from petroleum systems and stationary combustion. Methane emissions from coal mines dropped by 30 percent (29.5 $MMTCO_2e$) between 1990 and 2004. This decline resulted partly from the increased capture and use of methane from coal mine degasification systems and a shift in production away from some of the Nation's gassiest underground mines in Central Appalachia. Also, between 1990 and 2004, the share of coal production represented by underground mines declined from 41 percent to 34 percent. Methane emissions from underground mines tend to be higher than emissions from surface mines per ton of coal mined, because coal mined from the surface has been subjected to lower pressures and methane in the seams of surface mines has had earlier opportunities to migrate to the surface through cracks and fissures.

Methane emissions from petroleum systems dropped from 29.9 $MMTCO_2e$ in 1990 to 23.2 $MMTCO_2e$ in 2004.

A decrease of 5.0 $MMTCO_2e$ in estimated methane emissions from stationary combustion (from 13.0 $MMTCO_2e$ in 1990 to 8.0 $MMTCO_2e$ in 2004) made a smaller contribution to the overall drop in emissions from energy sources between 1990 and 2004. Together, the declines in emissions from coal mining, petroleum systems, and stationary combustion more than compensated for the increase of 23.7 $MMTCO_2e$ in emissions from natural gas systems, attributed to increasing U.S. consumption of natural gas between 1990 and 2004.

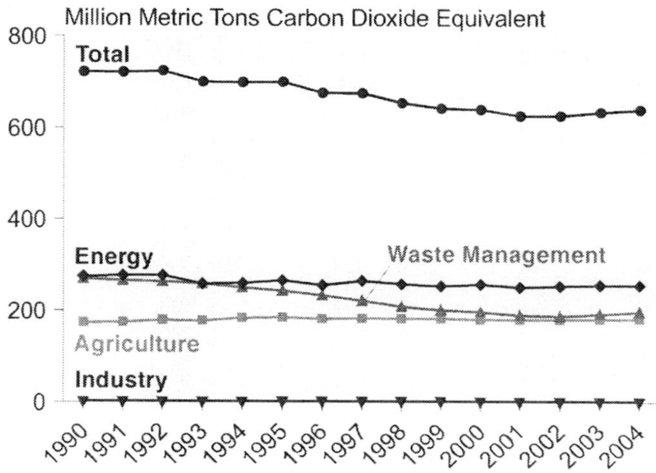

Source: Estimates presented in this chapter.

Figure 2. U.S. Emissions of Methane by Source, 1990-2004

Coal Mining

U.S. Methane Emissions from Coal Mining, 1990-2004

	Methane	Carbon Dioxide Equivalent
Estimated 2004 Emissions (Million Metric Tons)	3.0	68.2
Change Compared to 2003 (Million Metric Tons)	*	1.0
Change from 2003 *(Percent)*	*1.5%*	*1.5%*
Change Compared to 1990 (Million Metric Tons)	-1.3	-29.5
Change from 1990 *(Percent)*	*-30.2%*	*-30.2%*

*Less than 0.05 million metric tons.

The preliminary estimate of methane emissions from coal mines for 2004 is 68.2 MMTCO$_2$e (Table 16), an increase of 1.5 percent (1.0 MMTCO$_2$e) from the 2003 level of 67.2 MMTCO$_2$e. This increase can be traced primarily to a 3.7percent increase in coal production in 2004, centered predominantly in underground mines (up by 6.8 percent). U.S. coal production rose to 1.11 billion short tons in 2004, up from 1.07 billion short tons in 2003. The increase in coal production resulted from robust economic growth in 2004 that was accompanied by higher demand for coal to produce electricity and by record levels of coal exports.[65]

Methane emissions from coal mines have dropped by 30 percent, from 97.7 MMTCO$_2$e in 1990 to 68.2 MMTCO$_2$e in 2004. The decline is attributed to three important trends: (1)

methane recovery from active coal mines for use as an energy resource increased from 6.1 $MMTCO_2e$ in 1990 to about 15.2. $MMTCO_2e$ in 2004; (2) methane emissions from degasification systems were reduced by nearly 8.5 $MMTCO_2e$, from 28.9 $MMTCO_2e$ in 1990 to 20.4 $MMTCO_2e$ in 2004; and (3) decreases in coal production from gassy mines, combined with enhanced methane recovery though degasification, caused methane emissions from ventilation systems at gassy mines to drop by about 12.8 $MMTCO_2e$, from 48.9 $MMTCO_2e$ in 1990 to 36.1 $MMTCO_2e$ in 2004 (Table 16).[66]

Abandoned coal mines represent a significant source of additional emissions that has not been incorporated into the overall estimate of methane emissions in this report because of uncertainties associated with the data. The text box on page 39 provides a discussion of those uncertainties and the potential magnitude of additional emissions.

Natural Gas Systems

U.S. Methane Emissions from Natural Gas Systems, 1990-2004

	Methane	Carbon Dioxide Equivalent
Estimated 2004 Emissions (Million Metric Tons)	6.6	152.6
Change Compared to 2003 (Million Metric Tons)	*	-0.5
Change from 2003 *(Percent)*	*-0.3%*	*-0.3%*
Change Compared to 1990 (Million Metric Tons)	1.0	23.7
Change from 1990 *(Percent)*	*18.4%*	*18.4%*

*Less than 0.05 million metric tons.

Petroleum Systems

U.S. Methane Emissions from Petroleum Systems, 1990-2004

	Methane	Carbon Dioxide Equivalent
Estimated 2004 Emissions (Million Metric Tons)	1.0	23.2
Change Compared to 2003 (Million Metric Tons)	*	-0.1
Change from 2003 *(Percent)*	*-0.4%*	*-0.4%*
Change Compared to 1990 (Million Metric Tons)	-0.3	-6.7
Change from 1990 *(Percent)*	*-22.5%*	*-22.5%*

*Less than 0.05 million metric tons.

At 152.6 $MMTCO_2e$, 2004 estimated methane emissions from natural gas systems were down by 0.3 percent from the 153.1 $MMTCO_2e$ emitted in 2003 (Table 17). The 2004 estimate is preliminary, however, because pipeline data for 2004 have not been finalized as of the publication of this report. The estimated 2004 emissions level is 18 percent (23.7 $MMTCO_2e$) above the 1990 level (128.9 $MMTCO_2e$), with more than threefifths of the increase attributable to increased mileage of transmission and distribution pipelines and almost twofifths attributable to increases in natural gas production.[67]

Methane emissions from petroleum systems in 2004 are estimated at 23.2 $MMTCO_2e$, nearly unchanged from their 2003 level and down by 22 percent (6.7 $MMTCO_2e$) from their 1990 level of 29.9 $MMTCO_2e$. The decline in emissions from this source is almost exclusively due to a 26percent reduction in domestic oil production from 1990 to 2004. Approximately 92 percent (21.4 $MMTCO_2e$) of all U.S. emissions from petroleum systems occur during oil exploration and production (Table 18). A much smaller portion of methane emissions from petroleum systems can be traced to refineries (0.6 $MMTCO_2e$) and transportation of crude oil (1.1 $MMTCO_2e$).

Stationary Combustion

U.S. Methane Emissions from Stationary Combustion, 1990-2004

	Methane	Carbon Dioxide Equivalent
Estimated 2004 Emissions (Million Metric Tons)	0.3	8.0
Change Compared to 2003 (Million Metric Tons)	*	-0.5
Change from 2003 *(Percent)*	*-6.3%*	*-6.3%*
Change Compared to 1990 (Million Metric Tons)	-0.2	-5.0
Change from 1990 *(Percent)*	*-38.7%*	*-38.7%*

*Less than 0.05 million metric tons.

U.S. methane emissions from stationary combustion in 2004 were 8.0 $MMTCO_2e$, down by 6.3 percent from their 2003 level of 8.5 $MMTCO_2e$ and 39 percent below their 1990 level of 13.0 $MMTCO_2e$ (Table 19). Residential wood consumption typically accounts for about 85 percent of methane emissions from stationary combustion. Methane emissions are the result of incomplete combustion, and residential woodstoves and fireplaces provide much less efficient combustion than industrial or utility boilers. Estimates of emissions from residential wood combustion have fallen by 43 percent, from 11.8 $MMTCO_2e$ in 1990 to 6.7 $MMTCO_2e$ in 2004, although these estimates are very uncertain.[68]

The universe of wood consumers is large and heterogeneous, and the Energy Information Administration (EIA) collects data on residential wood consumption only at 4year intervals in its Residential Energy Consumption Survey (RECS). The most recently published EIA data on residential wood consumption are from the 2001 RECS.[69] Updated data on residential wood consumption for calendar year 2004 will be available from the 2005 RECS.

Mobile Combustion

Estimated U.S. methane emissions from mobile combustion in 2004 were 4.4 $MMTCO_2e$, up by 0.7 percent from the 2003 level but 22 percent lower than the 1990 level of 5.6 $MMTCO_2e$ (Table 20). Methane emissions from passenger cars have declined since 1990 as older vehicles with catalytic converters that are less efficient at destroying methane have been taken off the road. Estimates of methane emissions from mobile sources have been revised downward in the last two annual editions of this report, reflecting a change in the source of data for vehicle miles traveled and a related adjustment in the emission factors for lightduty trucks.[70]

Waste Management

U.S. Methane Emissions from Mobile Sources, **1990-2004**

	Methane	Carbon Dioxide Equivalent
Estimated 2004 Emissions (Million Metric Tons)	0.2	4.4
Change Compared to 2003 (Million Metric Tons)	*	*
Change from 2003 *(Percent)*	*0.7%*	*0.7%*
Change Compared to 1990 (Million Metric Tons)	-0.1	-1.2
Change from 1990 *(Percent)*	*-22.1%*	*-22.1%*

*Less than 0.05 million metric tons.

Methane emissions from waste management, at 198.2 $MMTCO_2e$, accounted for 31 percent of U.S. anthropogenic methane emissions in 2004 (Figure 2). Emissions from this source have fallen by 27 percent (72.0 $MMTCO_2e$) from their 1990 level of 270.2 $MMTCO_2e$. Landfills represent 92 percent (182.6 $MMTCO_2e$) of the methane emissions from waste management in 2004, and they are the largest single source of U.S. anthropogenic methane emissions (Table 15). The remainder of emissions from waste management (15.6 $MMTCO_2e$) is associated with domestic wastewater treatment. Estimated emissions from waste management would increase if sufficient information were available to develop a reliable estimate of emissions from industrial wastewater treatment.

EIA's estimates of methane emissions from landfills include emissions from both municipal solid waste landfills and industrial landfills. Estimated methane emissions from industrial landfills are based on a methodology developed by the EPA and represent 7 percent of emissions from municipal solid waste landfills.[71]

EIA has revised its estimates of methane recovered for energy from landfills. Previous editions of this report erroneously included the avoided emissions of carbon dioxide from fossil fuel combustion displaced by landfill gastoenergy operations in the estimate of methane recovered. This resulted in double counting of the impacts of fossil fuel displacement by landfill gastoenergy. For this year's report, the avoided emissions have been removed from estimates going back to 1990. Similarly, estimates of methane recovered and flared have been lowered to eliminate some potential double counting associated with methane recovery for energy projects, as flares are typically used for backup systems at plants that recover methane for energy. Together, these revisions have raised the EIA estimates of overall methane emissions from waste management by increments ranging from 8 MMTCO_2e for 1990 to 21 MMTCO_2e for 2003.

Methane Emissions from Abandoned Coal Mines

Thousands of coal mines in the United States have been closed and abandoned during the past 100 years. The U.S. Department of Labor's Mine Safety and Health Administration (MSHA) estimates that since 1980 more than 7,500 coal mines have been abandoned, and many continue to emit methane. In an April 2004 report,[a] the U.S. Environmental Protection Agency (EPA) estimated that methane emissions from abandoned coal mines ranged between 3.0 MMTCO_2e and 4.6MMTCO_2e in 1990, and between 4.6MMTCO_2e and 6.4 MMTCO_2e in 2002. More recently, the EPA estimated methane emissions of 7.0MMTCO_2e from abandoned underground coal mines in 2003, up from 6.7 MMTCO_2e in 1990 but down from a peak of 8.4 MMTCO_2e in 2000 due to a decline in the number of gassy mines being abandoned.[b] Because access to abandoned mines is limited and a systematic measurement program at those sites would be timeintensive and costly, the EPA estimates rely on actual emissions data from when the mines were operating and assume a decline function in emissions based on mine and coalseam characteristics.

The most important variable in determining the amount of methane emissions from an abandoned mine is its postmining status—whether the mine has been sealed, flooded, or continues to be vented. Sealed and flooded mines have much lower rates of emissions than vented mines. Another variable deemed important is whether the mine was gassy (emitting more than 100,000 cubic feet per day) when it was operating. Gassy mines are estimated to emit 98 percent of all methane emissions from operating coal mines, and the EPA assumes that abandoned mines which had been gassy when operating represent a similarly predominant portion of emissions from abandoned mines. The EPA's 2004 study thus focuses on abandoned mines that had been gassy prior to closure. Of the 364 gassy mines abandoned since 1972, the EPA has data on the status of a portion of them (i.e., whether the mines were sealed, flooded, or continue to be vented), calculates percentage shares of emissions by their respective status, and then assumes that those shares apply to mines for which it does not have data.

For abandoned mines that have been vented, the EPA derives an emissions decline curve based on three primary factors: adsorption isotherms by coal basin, coal permeability estimates, and estimates of pressure at abandonment. For mines that are flooded, the EPA assumes a decline curve equation based on measurements taken from eight abandoned mines in two basins. The EPA treats sealed mines similarly to those vented, adjusting the initial emissions rate and length of time for emissions to dissipate, given the slower release rate from sealed vents.

The EPA sought to calibrate its estimation methodology to field measurements, but restricted access precluded measurement at all but seven mines. Although results from those mines suggested the general accuracy of the estimation method, EIA believes the methodology has not yet been validated. EIA will reconsider including estimates of methane emissions from abandoned mines in its overall estimates of U.S. greenhouse gas emissions should additional field data confirm the EPA methodology.

Landfills

U.S. Methane Emissions from Landfills, 1990-2004

	Methane	Carbon Dioxide Equivalent
Estimated 2004 Emissions (Million Metric Tons)	7.9	182.6
Change Compared to 2003 (Million Metric Tons)	0.2	4.5
Change from 2003 *(Percent)*	*2.5%*	*2.5%*
Change Compared to 1990 (Million Metric Tons)	-3.2	-74.4
Change from 1990 *(Percent)*	*-29.0%*	*-29.0%*

Due to increased levels of waste disposed in landfills, estimated methane emissions from landfills rose to 182.6 MMTCO$_2$e in 2004, 2.5 percent (4.5 MMTCO$_2$e) above the 2003 level of 178.1 MMTCO$_2$e but still 29 percent (74.4 MMTCO$_2$e) below the 1990 level of 257.0 MMTCO$_2$e (Table 21). The dramatic decrease in methane emissions since 1990 is directly attributable to a 100.4 MMTCO$_2$e increase in methane captured at landfills that otherwise would have been emitted to the atmosphere. Of the 122.1 MMTCO$_2$e of methane believed to be captured from this source in 2004, 63.3 MMTCO$_2$e was recovered for energy use, and 58.9 MMTCO$_2$e was recovered and flared. In 2004, methane recovery for energy increasingly took the form of direct use of mediumBtu gas in industrial boilers. The acceleration of this practice was driven by higher natural gas prices, which made landfill gas more competitive.[72]

Estimates of methane recovered for energy are drawn from data collected by the U.S. Environmental Protection Agency's (EPA's) Landfill Methane Outreach Program.[73] Estimates of methane recovered and flared are based on data collected from flaring equipment vendors, in conjunction with data reported on Form EIA1605.[74] There is less uncertainty in the estimate of methane recovered and used for energy, and it is likely that estimates of methane flared are biased downward due to a lack of comprehensive industry data.

The rapid growth in methane recovery has been aided by a combination of regulatory and tax policy. The Federal Section 29 (of the Internal Revenue Code) tax credit for alternative energy sources, added to the tax code as part of the Crude Oil Windfall Profits Act of 1980, provided a subsidy roughly equivalent to 1 cent per kilowatthour for electricity generated from landfill gas; however, the tax credit expired on June 30, 1998. As part of the American Jobs

Creation Act of 2004, a tax credit for electricity generation using landfill gas was added to Section 45 of the Internal Revenue Code. The credit was augmented under the Energy Policy Act of 2005, which extended the credit period—previously 5 years from the original date of service—to 10 years from the original date of service. To be eligible for the credit, a landfill gastoenergy project must be placed in service between October 22,2004, and December 31,2007. Those facilities that qualify are eligible for a 5year tax credit valued at 0.9 cent per kilowatthour.

Increases in methane recovery have also resulted from the implementation of the EPA's New Source Performance Standards and Emission Guidelines. The regulations require all landfills with more than 2.5 million metric tons of waste in place and annual emissions of nonmethane organic compounds (NMOCs) exceeding 50 metric tons to collect and burn their landfill gas, either by flaring or for use as an energy source.

Domestic and Commercial Wastewater Treatment

With the U.S. population growing slowly, methane emissions from domestic and commercial wastewater treatment are estimated to have grown by nearly 1.0 percent between 2003 and 2004 to 15.6 MMTCO$_2$e—about 18 percent above the 1990 level of 13.2 MMTCO$_2$e (Table 15). Methane emissions from industrial wastewater treatment are discussed in the text box on page 41.

U.S. Methane Emissions from Domestic and Commercial Wastewater Treatment, 1990-2004

	Methane	Carbon Dioxide Equivalent
Estimated 2004 Emissions (Million Metric Tons)	0.7	15.6
Change Compared to 2003 (Million Metric Tons)	*	0.2
Change from 2003 *(Percent)*	*1.0%*	*1.0%*
Change Compared to 1990 (Million Metric Tons)	0.1	2.4
Change from 1990 *(Percent)*	*18.0%*	*18.0%*

*Less than 0.05 million metric tons.

Methane emissions from domestic and commercial wastewater treatment are a function of the share of organic matter in the wastewater stream and the conditions under which it decomposes. Wastewater may be treated aerobically or anaerobically. Because aerobic decomposition does not yield methane, whereas anaerobic decomposition does, the method of treatment is a critical determinant of emissions; however, there is little information available on wastewater treatment methods. Data on flaring or energy recovery from methane generated by wastewater are also sparse. EIA believes that emissions from this source are relatively small, representing less than 3 percent of all U.S. methane emissions in 2004. Thus, emissions are estimated using a default percapita emissions factor and U.S. population data.

Methane Emissions from Industrial Wastewater Treatment

When wastewater containing large amounts of organic material is treated through anaerobic decomposition, methane is emitted. The best estimate of those emissions would be based on a systematic measurement of all point sources; however, the number and diversity of U.S. industrial wastewater sources make such an approach unaffordable and impractical. As an alternative, methane emissions from industrial wastewater treatment are calculated by the following equation:

$$M = PxOxCODxAxEF\ ,$$

where M = methane emissions, P = product output, O = wastewater outflow per unit of product output, COD = organic loading in outflow, A = percentage of outflow treated anaerobically, and EF = emissions factor for anaerobically treated outflow.

The Intergovernmental Panel on Climate Change (IPCC), in its Good Practice Guidance and Uncertainty Management in National Greenhouse Gas Inventories [a], provides default data for wastewater generation and COD on an industryspecific basis. The default data, often based on one or two literature sources, are assumed to have an uncertainty range of minus 50 percent to plus 100 percent (although no justification for the range is provided). The IPCC also provides a single default factor of 0.25 kilograms methane per kilogram of COD, premised on a general approximation of the theoretical maximum for this emission factor, and identifies an uncertainty of plus or minus 30 percent for this estimate.

There are currently no specific U.S. data that could be used to improve on the IPCC defaults, and the uncertainties make it impossible for the Energy Information Administration to provide a reliable estimate of emissions from this source. It can be noted, however, that depending on the extent to which industrial wastewater from such industries as meat and poultry processing, pulp and paper manufacturing, and vegetable, fruit, and juice processing—which is likely to have a high content of organic material—is treated anaerobically, excluding the methane emissions that would result from the U.S. emissions total will tend to produce an underestimate of U.S. methane emissions. The U.S. Environmental Protection Agency estimates that U.S. methane emissions from industrial wastewater treatment could be as high as 16.9 MMTCO2e in 2003[b]. If more comprehensive data on industrial wastewater flows become available, EIA will consider adding this source to its estimate of U.S. methane emissions.

Agricultural Sources

Estimated methane emissions from agricultural sources, at 182.3 $MMTCO_2e$ in 2004, represent 28.5 percent of total U.S. anthropogenic methane emissions (Table 15). Agricultural methane emissions increased by less than 1 percent (0.9 $MMTCO_2e$) from 2003 to 2004, as a small decrease in emissions from enteric fermentation was offset by increases in emissions from animal waste management, rice cultivation, and crop residue burning. Of total estimated methane emissions from agricultural activities, 93 percent (170.0 $MMTCO_2e$) results from livestock management, of which 67.8 percent (115.2 $MMTCO_2e$) can be traced to enteric fermentation in ruminant animals and the remainder (54.7 $MMTCO_2e$) to anaerobic decomposition of livestock wastes. A small portion of U.S. agricultural methane emissions result from crop residue burning and wetland rice cultivation.

Enteric Fermentation in Domesticated Animals

U.S. Methane Emissions from Enteric Fermentation in Domesticated Animals, 1990-2004

	Methane	Carbon Dioxide Equivalent
Estimated 2004 Emissions (Million Metric Tons)	5.0	115.2
Change Compared to 2003 (Million Metric Tons)	*	-0.9
Change from 2003 *(Percent)*	*-0.8%*	*-0.8%*
Change Compared to 1990 (Million Metric Tons)	-0.2	-4.4
Change from 1990 *(Percent)*	*-3.7%*	*-3.7%*

*Less than 0.05 million metric tons.

In 2004, estimated methane emissions from enteric fermentation in domesticated animals declined by 0.8 percent to 115.2 $MMTCO_2e$ (Table 22). Because 95 percent of all emissions from enteric fermentation is attributable to cattle (109.9 $MMTCO_2e$), trends in emissions correlate with trends in cattle populations. Between 2003 and 2004, cattle populations were nearly constant, with small declines in all population categories offset somewhat by increases in populations of beef cattle on feedlots. Estimated methane emissions from enteric fermentation in 2004 are 3.7 percent below their 1990 level of 119.6 $MMTCO_2e$.

Solid Waste of Domesticated Animals

Estimated methane emissions from the solid waste of domesticated animals increased from 54.2 $MMTCO_2e$ in 2003 to 54.7 $MMTCO_2e$ in 2004 (Table 23). The increase reinforced a larger trend over the past decade: in 2004, emissions from the solid waste of domesticated animals were 11.2 $MMTCO_2e$ above their 1990 level of 43.5 $MMTCO_2e$, an increase of 26 percent. Between 1990 and 2004, there was a shift in livestock management to larger facilities, which are believed to be more likely to manage waste using liquid systems that tend to promote methane generation.[75]

Rice Cultivation

Estimated methane emissions from U.S. rice cultivation increased to 11.0 $MMTCO_2e$ in 2004 from 9.8 $MMTCO_2e$ in 2003 (Table 15). The rise was the result of a 12percent jump in the number of acres harvested.[76] All U.S. riceproducing States saw increases in acres harvested during 2004, and total methane emissions from rice cultivation in 2004 were 18 percent (1.7 MMTCO_2e) higher than in 1990.

Burning of Crop Residues

Crop residue burning, the smallest contributor to U.S. methane emissions, represents less than 1 percent of total U.S. methane emissions. Estimated 2004 methane emissions from the burning of crop residues were 1.3 MMTCO_2e, up by 12 percent from 2003 and 27 percent above their 1990 level of 1.0 MMTCO_2e (Table 15). The increase from 2003 to 2004 is attributable mainly to large increases in corn and soybean production.

U.S. Methane Emissions from Solid Waste of Domesticated Animals, 1990-2004

	Methane	Carbon Dioxide Equivalent
Estimated 2004 Emissions (Million Metric Tons)	2.4	54.7
Change Compared to 2003 (Million Metric Tons)	*	0.5
Change from 2003 *(Percent)*	*0.9%*	*0.9%*
Change Compared to 1990 (Million Metric Tons)	0.5	11.2
Change from 1990 *(Percent)*	*25.8%*	*25.8%*

*Less than 0.05 million metric tons.

Industrial Processes

U.S. Methane Emissions from Industrial Processes, 1990-2004

	Methane	Carbon Dioxide Equivalent
Estimated 2004 Emissions (Million Metric Tons)	0.1	2.7
Change Compared to 2003 (Million Metric Tons)	*	0.1
Change from 2003 *(Percent)*	*4.1%*	*4.1%*
Change Compared to 1990 (Million Metric Tons)	*	*
Change from 1990 *(Percent)*	*0.2%*	*0.2%*

*Less than 0.05 million metric tons.

Table 15. U.S. Methane Emissions from Anthropogenic Sources, **1990 and 1996-2004**

Source	1990	1996	1997	1998	1999	2000	2001	2002	2003	P2004
	Million Metric Tons Carbon Dioxide Equivalent									
Energy Sources										
Coal Mining	97.7	73.4	80.4	75.6	71.7	68.5	68.0	64.7	67.2	68.2
Natural Gas Systems	128.9	138.1	144.3	143.9	144.2	151.0	147.0	154.0	153.1	152.6
Petroleum Systems	29.9	26.4	26.3	25.5	24.0	23.8	23.7	23.5	23.3	23.2
Stationary Combustion	13.0	13.3	10.1	9.1	9.6	10.1	8.7	7.6	8.5	8.0
Mobile Sources	5.6	5.1	5.1	4.8	4.8	4.7	4.6	4.6	4.3	4.4
Total Energy Sources	**275.0**	**256.3**	**266.2**	**258.9**	**254.4**	**258.1**	**252.0**	**254.3**	**256.4**	**256.3**
Waste Management										
Landfills	257.0	219.8	208.1	195.0	187.4	183.3	175.8	173.4	178.1	182.6
Wastewater Treatment	13.2	14.3	14.5	14.7	14.8	15.0	15.2	15.3	15.5	15.6
Total Waste Management	**270.2**	**234.2**	**222.6**	**209.7**	**202.2**	**198.3**	**191.0**	**188.7**	**193.6**	**198.2**
Agricultural Sources										
Enteric Fermentation	119.6	122.0	119.1	117.2	117.3	116.3	115.1	115.7	116.2	115.2
Animal Waste	43.5	49.7	52.7	53.6	52.7	52.8	53.3	53.7	54.2	54.7
Rice Cultivation	9.3	9.4	10.3	10.7	11.5	10.2	10.7	10.2	9.8	11.0
Crop Residue Burning	1.0	1.1	1.1	1.1	1.1	1.1	1.1	1.0	1.2	1.3
Total Agricultural Sources	**173.4**	**182.2**	**183.3**	**182.6**	**182.5**	**180.5**	**180.3**	**180.6**	**181.3**	**182.3**
Industrial Processes	**2.7**	**3.1**	**3.1**	**3.1**	**3.1**	**2.9**	**2.5**	**2.6**	**2.6**	**2.7**
Total	**721.4**	**675.8**	**675.2**	**654.2**	**642.2**	**639.8**	**625.8**	**626.2**	**633.9**	**639.5**
	Million Metric Tons Methane									
Energy Sources										
Coal Mining	4.25	3.19	3.50	3.29	3.12	2.98	2.96	2.81	2.92	2.96
Natural Gas Systems	5.60	6.00	6.27	6.26	6.27	6.57	6.39	6.70	6.66	6.64
Petroleum Systems	1.30	1.15	1.14	1.11	1.04	1.03	1.03	1.02	1.01	1.01
Stationary Combustion	0.56	0.58	0.44	0.39	0.42	0.44	0.38	0.33	0.37	0.35
Mobile Sources	0.24	0.22	0.22	0.21	0.21	0.21	0.20	0.20	0.19	0.19
Total Energy Sources	**11.96**	**11.14**	**11.57**	**11.25**	**11.06**	**11.22**	**10.96**	**11.06**	**11.15**	**11.14**
Waste Management										
Landfills	11.17	9.56	9.05	8.48	8.15	7.97	7.64	7.54	7.74	7.94
Wastewater Treatment	0.58	0.62	0.63	0.64	0.65	0.65	0.66	0.67	0.67	0.68
Total Waste Management	**11.75**	**10.18**	**9.68**	**9.12**	**8.79**	**8.62**	**8.30**	**8.20**	**8.42**	**8.62**
Agricultural Sources										
Enteric Fermentation	5.20	5.30	5.18	5.10	5.10	5.06	5.01	5.03	5.05	5.01
Animal Waste	1.89	2.16	2.29	2.33	2.29	2.29	2.32	2.33	2.36	2.38
Rice Cultivation	0.40	0.41	0.45	0.47	0.50	0.44	0.47	0.45	0.43	0.48
Crop Residue Burning	0.04	0.05	0.05	0.05	0.05	0.05	0.05	0.05	0.05	0.06
Total Agricultural Sources	**7.54**	**7.92**	**7.97**	**7.94**	**7.94**	**7.85**	**7.84**	**7.85**	**7.88**	**7.92**
Industrial Processes	**0.12**	**0.13**	**0.13**	**0.13**	**0.13**	**0.13**	**0.11**	**0.11**	**0.11**	**0.12**
Total	**31.36**	**29.38**	**29.36**	**28.44**	**27.92**	**27.82**	**27.21**	**27.23**	**27.56**	**27.80**

P = preliminary data.

Notes: Data in this table are revised from the data contained in the previous EIA report, *Emissions of Greenhouse Gases in the United States 2003,* DOE/EIA0573(2003) (Washington, DC, December 2004). Totals may not equal sum of components due to independent rounding.

Sources: EIA estimates presented in this chapter. Emissions calculations based on Intergovernmental Panel on Climate Change, *Greenhouse Gas Inventory Reference Manual: Revised 1996 IPCC Guidelines for National Greenhouse Gas Inventories,* Vol. 3 (Paris, France, 1997), pp. 4.834.84, web site www.ipcc.ch/pub/guide.htm; and U.S. U.S. Environmental Protection Agency, Inventory of U.S. Greenhouse Gas Emissions and Sinks 19902003, EPA430R05003 (Washington, DC, April 2005), web site http://yosemite.epa.gov/oar/globalwarming.nsf/content/Resource CenterPublicationsGHGEmissions USEmissionsInventory2005.html.

Chemical Production

The preliminary estimate of methane emissions from U.S. chemical production in 2004 is 1.6 $MMTCO_2e$, up by 6 percent from the 2003 level of 1.5 $MMTCO_2e$. Methane emissions from chemical production in 2004 were 25 percent above their 1990 level of 1.3 $MMTCO_2e$.

The increase is attributable to increased production of carbon black, ethylene, and styrene, which more than offset a drop in methanol production (Table 24).[77]

Iron and Steel Production

With production of pig iron rebounding from a 2decade low in 2003,[78] methane emissions from iron and steel production rose to an estimated 1.1 $MMTCO_2e$ in 2004, a 2.2percent increase from 2003 but still 23 percent below their 1990 level of 1.4 $MMTCO_2e$ (Table 24).

Table 16. U.S. Methane Emissions from Coal Mining and PostMining Activities, 1990 and 1996-2004

Source	1990	1996	1997	1998	1999	2000	2001	2002	2003	P2004
Million Metric Tons Carbon Dioxide Equivalent										
Surface Mining										
Mining	9.8	10.6	10.9	11.4	11.5	11.4	12.1	12.0	11.7	11.9
Post-Mining	0.9	0.9	0.9	1.0	1.0	1.0	1.1	1.0	1.0	1.0
Subtotal	*10.7*	*11.5*	*11.8*	*12.4*	*12.5*	*12.4*	*13.2*	*13.0*	*12.7*	*13.0*
Underground Mining										
Ventilation (Gassy Mines)[a]	48.9	39.3	41.2	41.5	40.6	38.3	37.2	34.7	33.4	36.1
Ventilation (Nongassy Mines)	0.6	0.8	0.8	0.9	0.9	0.9	0.8	0.8	0.8	0.8
Degasification Systems	28.9	23.5	24.4	21.8	18.1	20.0	21.3	22.9	24.6	20.4
Post-Mining	14.7	14.2	14.6	14.5	13.6	12.9	13.2	12.4	12.2	13.0
Methane Recovery for Energy (-)	6.1	16.0	12.5	15.5	13.9	16.0	17.7	19.1	16.5	15.2
Subtotal	*87.0*	*61.9*	*68.6*	*63.2*	*59.2*	*56.1*	*54.8*	*51.7*	*54.5*	*55.2*
Net Emissions	**97.7**	**73.4**	**80.4**	**75.6**	**71.7**	**68.5**	**68.0**	**64.7**	**67.2**	**68.2**
Million Metric Tons Methane										
Surface Mining										
Mining	0.43	0.46	0.47	0.49	0.50	0.49	0.53	0.52	0.51	0.52
Post-Mining	0.04	0.04	0.04	0.04	0.04	0.04	0.05	0.05	0.04	0.05
Subtotal	*0.46*	*0.50*	*0.51*	*0.54*	*0.54*	*0.54*	*0.57*	*0.57*	*0.55*	*0.56*
Underground Mining										
Ventilation (Gassy Mines)[a]	2.13	1.71	1.79	1.80	1.76	1.67	1.62	1.51	1.45	1.57
Ventilation (Nongassy Mines)	0.03	0.04	0.04	0.04	0.04	0.04	0.04	0.03	0.03	0.04
Degasification Systems	1.26	1.02	1.06	0.95	0.79	0.87	0.93	1.00	1.07	0.89
Post-Mining	0.64	0.62	0.63	0.63	0.59	0.56	0.57	0.54	0.53	0.57
Methane Recovery for Energy (-)	0.26	0.69	0.54	0.67	0.60	0.70	0.77	0.83	0.72	0.66
Subtotal	*3.78*	*2.69*	*2.98*	*2.75*	*2.57*	*2.44*	*2.38*	*2.25*	*2.37*	*2.40*
Net Emissions	**4.25**	**3.19**	**3.50**	**3.29**	**3.12**	**2.98**	**2.96**	**2.81**	**2.92**	**2.96**

[a]A "gassy" mine is an underground mine with ventilation emissions of 100,000 cubic feet of methane or more per day, as measured by the U.S. Mine Safety and Health Administration.

P = preliminary data.

Notes: Data in this table are revised from the data contained in the previous EIA report, Emissions of Greenhouse Gases in the United States 2003, DOE/EIA0573(2003) (Washington, DC, December 2004). Totals may not equal sum of components due to independent rounding.

Sources: Coal production numbers from Energy Information Administration, Coal Production, DOE/EIA0118 (Washington, DC, various years), and Coal Industry Annual, DOE/EIA0584 (Washington, DC, various years). Methane recovery rates from U.S. Environmental Protection Agency, Office of Air and Radiation, NonCO_2 Gases and Sequestration Branch, Coalbed Methane Outreach Program. Ventilation data for 1985, 1988, and 1990 provided by G. Finfinger, U.S. Department of the Interior, Bureau of Mines, Pittsburgh Research Center. Ventilation data for all other years provided by U.S. Environmental Protection Agency, Office of Air and Radiation, NonCO_2 Gases and Sequestration Branch, Coalbed Methane Outreach Program.

Table 17. U.S. Methane Emissions from Natural Gas Systems, 1990 and 1996-2004

Source	1990	1996	1997	1998	1999	2000	2001	2002	2003	P2004
Million Metric Tons Carbon Dioxide Equivalent										
Production	33.8	36.3	38.0	38.4	37.3	39.9	42.0	42.4	42.9	42.8
Natural Gas Processing	14.9	16.9	16.4	16.0	16.1	16.4	16.0	15.5	14.4	14.4
Transmission and Storage	48.3	48.5	52.1	51.0	52.8	55.4	49.0	55.6	54.7	54.3
Distribution	32.0	36.3	37.8	38.5	38.0	39.3	40.0	40.5	41.2	41.2
Total	**128.9**	**138.1**	**144.3**	**143.9**	**144.2**	**151.0**	**147.0**	**154.0**	**153.1**	**152.6**
Natural Gas STAR Reductions	0.3	8.3	10.7	12.9	14.5	16.5	19.2	25.9	25.9	27.0
Million Metric Tons Methane										
Production	1.47	1.58	1.65	1.67	1.62	1.73	1.83	1.84	1.86	1.86
Natural Gas Processing	0.65	0.73	0.71	0.69	0.70	0.71	0.69	0.67	0.63	0.63
Transmission and Storage	2.10	2.11	2.27	2.22	2.30	2.41	2.13	2.42	2.38	2.36
Distribution	1.39	1.58	1.64	1.67	1.65	1.71	1.74	1.76	1.79	1.79
Total	**5.60**	**6.00**	**6.27**	**6.26**	**6.27**	**6.57**	**6.39**	**6.70**	**6.66**	**6.64**
Natural Gas STAR Reductions	0.01	0.36	0.47	0.56	0.63	0.72	0.84	1.13	1.13	1.17

P = preliminary data.

Notes: Data in this table are revised from the data contained in the previous EIA report, *Emissions of Greenhouse Gases in the United States 2003,* DOE/EIA0573(2003) (Washington, DC, December 2004). Totals may not equal sum of components due to independent rounding. Data for Natural Gas STAR reductions are estimates provided by the EPA, based on annual submissions to the EPA by companies participating in the program, which report activities undertaken to avoid methane emissions from natural gas and petroleum systems.

Sources: National Risk Management Research Laboratory, *Methane Emissions From the Natural Gas Industry,* Vol. 2, Technical Report, GRI94/0257.1 and EPA600R9608 (Research Triangle Park, NC, June 1996), Appendix A; American Gas Association, *Gas Facts* (various years); Energy Information Administration, *Natural Gas Annual,* DOE/EIA01 31 (various years); Energy Information Administration, *Monthly EnergyReview,* DOE/EIA0035(2004/07) (Washington, DC, July 2004); Energy Information Administration, *Petroleum SupplyAnnual,* DOE/EIA0340 (Washington, DC, various years).

Table 18. U.S. Methane Emissions from Petroleum Systems, 1990 and 1996-2004

Source	1990	1996	1997	1998	1999	2000	2001	2002	2003	P2004
Million Metric Tons Carbon Dioxide Equivalent										
Refineries	0.6	0.6	0.6	0.6	0.6	0.6	0.6	0.6	0.6	0.6
Exploration and Production	29.0	25.5	25.5	24.7	23.2	22.9	22.9	22.6	22.4	21.4
Crude Oil Transportation	0.3	0.3	0.3	0.3	0.2	0.2	0.2	0.2	0.2	1.1
Total	**29.9**	**26.4**	**26.3**	**25.5**	**24.0**	**23.8**	**23.7**	**23.5**	**23.3**	**23.2**
Million Metric Tons Methane										
Refineries	0.02	0.03	0.03	0.03	0.03	0.03	0.03	0.03	0.03	0.03
Exploration and Production	1.26	1.11	1.11	1.07	1.01	1.00	0.99	0.98	0.97	0.93
Crude Oil Transportation	0.01	0.01	0.01	0.01	0.01	0.01	0.01	0.01	0.01	0.05
Total	**1.30**	**1.15**	**1.14**	**1.11**	**1.04**	**1.03**	**1.03**	**1.02**	**1.01**	**1.01**

P = preliminary data.

Notes: Data in this table are revised from the data contained in the previous EIA report, *Emissions of Greenhouse Gases in the United States 2003,* DOE/EIA0573(2003) (Washington, DC, December 2004). Totals may not equal sum of components due to independent rounding.

Sources: U.S. Environmental Protection Agency, Office of Air and Radiation, *Draft Estimates of Methane Emissions from the U.S. Oil Industry* (Draft Report, Washington, DC); Energy Information Administration, *Petroleum SupplyAnnual,* DOE/EIA0340 (Washington, DC, various years); and *Oil and Gas Journal,* Worldwide Refining Issue and Pipeline Economics Issue (various years).

Table 19. U.S. Methane Emissions from Stationary Combustion Sources, 1990 and 1996-2004

Source	1990	1996	1997	1998	1999	2000	2001	2002	2003	P2004
	Thousand Metric Tons Carbon Dioxide Equivalent									
Residential										
Coal	*	*	*	*	0	0	0	0	0	0
Fuel Oil[a]	113	107	101	89	96	105	105	99	105	109
Natural Gas	100	119	113	103	107	113	108	111	116	111
LPG	10	13	12	12	14	15	14	14	14	14
Wood	11,783	12,067	8,782	7,849	8,396	8,782	7,504	6,348	7,281	6,733
Total	**12,006**	**12,306**	**9,008**	**8,053**	**8,613**	**9,014**	**7,731**	**6,573**	**7,515**	**6,968**
Commercial										
Coal	29	28	30	23	24	20	21	21	19	20
Fuel Oil[a]	16	12	10	9	9	10	10	9	11	12
Natural Gas	71	86	88	82	83	86	83	86	88	82
LPG	2	2	2	2	3	3	3	3	2	3
Wood	3	3	3	3	4	4	3	3	3	3
Total	**120**	**131**	**133**	**120**	**121**	**122**	**119**	**121**	**123**	**119**
Industrial										
Coal	151	133	130	123	121	122	122	112	113	112
Fuel Oil[a]	32	27	24	20	18	20	21	17	19	23
Natural Gas	263	307	308	303	291	295	270	274	264	271
LPG	51	66	68	65	71	72	65	70	67	69
Wood	89	104	107	99	100	101	89	86	84	89
Total	**585**	**637**	**637**	**610**	**601**	**609**	**566**	**559**	**547**	**564**
Electric Power										
Coal	225	256	262	267	268	281	273	275	280	281
Fuel Oil[a]	19	10	12	17	16	14	16	11	14	14
Natural Gas	7	9	9	10	11	12	12	13	12	12
Wood	1	0	0	0	0	0	0	0	0	0
Total	**252**	**275**	**283**	**294**	**294**	**307**	**302**	**298**	**306**	**308**
Total All Sectors										
Coal	405	416	423	413	412	423	416	408	413	414
Fuel Oil[a]	180	156	147	135	139	149	153	137	149	158
Natural Gas	441	521	518	499	492	505	473	484	479	476
LPG	62	81	82	78	88	90	82	87	83	86
Wood	11,875	12,174	8,891	7,951	8,499	8,886	7,595	6,436	7,367	6,825
Total	**12,964**	**13,349**	**10,061**	**9,076**	**9,630**	**10,052**	**8,718**	**7,551**	**8,491**	**7,959**

*Less than 500 metric tons carbon dioxide equivalent.

[a]Fuel oil use in the residential sector consists of distillate fuel only. In the other sectors it includes both distillate and residual fuel oil.

P = preliminary data.

See notes and sources at end of table.

Table 19. U.S. Methane Emissions from Stationary Combustion Sources, **1990 and 1996-2004 (Continued)**

Source	1990	1996	1997	1998	1999	2000	2001	2002	2003	P2004
	Thousand Metric Tons Methane									
Residential										
Coal	*	*	*	*	0	0	0	0	0	0
Fuel Oil[a]	5	5	4	4	4	5	5	4	5	5
Natural Gas	4	5	5	4	5	5	5	5	5	5
LPG	*	1	1	1	1	1	1	1	1	1
Wood	512	525	382	341	365	382	326	276	317	293
Total	**522**	**535**	**392**	**350**	**374**	**392**	**336**	**286**	**327**	**303**
Commercial										
Coal	1	1	1	1	1	1	1	1	1	1
Fuel Oil[a]	1	1	*	*	*	*	*	*	*	1
Natural Gas	3	4	4	4	4	4	4	4	4	4
LPG	*	*	*	*	*	*	*	*	*	*
Wood	*	*	*	*	*	*	*	*	*	*
Total	**5**	**6**	**6**	**5**	**5**	**5**	**5**	**5**	**5**	**5**
Industrial										
Coal	7	6	6	5	5	5	5	5	5	5
Fuel Oil[a]	1	1	1	1	1	1	1	1	1	1
Natural Gas	11	13	13	13	13	13	12	12	11	12
LPG	2	3	3	3	3	3	3	3	3	3
Wood	4	5	5	4	4	4	4	4	4	4
Total	**25**	**28**	**28**	**27**	**26**	**26**	**25**	**24**	**24**	**25**
Electric Power										
Coal	10	11	11	12	12	12	12	12	12	12
Fuel Oil[a]	1	*	1	1	1	1	1	*	1	1
Natural Gas	*	*	*	*	*	1	1	1	1	1
Wood	*	0	0	0	0	0	0	0	0	0
Total	**11**	**12**	**12**	**13**	**13**	**13**	**13**	**13**	**13**	**13**
Total All Sectors										
Coal	18	18	18	18	18	18	18	18	18	18
Fuel Oil[a]	8	7	6	6	6	6	7	6	7	7
Natural Gas	19	23	23	22	21	22	21	21	21	21
LPG	3	4	4	3	4	4	4	4	4	4
Wood	516	529	387	346	370	386	330	280	320	297
Total	**564**	**580**	**437**	**395**	**419**	**437**	**379**	**328**	**369**	**346**

*Less than 500 metric tons methane.

[a]Fuel oil use in the residential sector consists of distillate fuel only. In the other sectors it includes both distillate and residual fuel oil.

P = preliminary data.

Notes: Data in this table are revised from the data contained in the previous EIA report, *Emissions of Greenhouse Gases in the United States 2003,* DOE/EIA0573(2003) (Washington, DC, December 2004). Totals may not equal sum of components due to independent rounding.

Sources: U.S. Environmental Protection Agency, Office of Air Quality Planning and Standards, *Compilation of Air Pollutant Emission Factors, Volume 1: StationaryPointandArea Sources,* AP 42, Fifth Edition (Washington, DC, January 1995), web site www.epa.gov/ttn/chief/ap42; Intergovernmental Panel on Climate Change, *Greenhouse Gas Inventory Reference Manual: Revised 1996 IPCC Guidelines for National Greenhouse Gas Inventories,* Vol. 3 (Paris, France, 1997), web site www.ipcc.ch/pub/guide.htm; and Energy Information Administration, *State Energy Data Report,* DOE/EIA0214 (Washington, DC, various years), *Monthly Energy Review,* DOE/EIA0035(2004/07) (Washington, DC, July 2004), and *Annual Energy Review,* DOE/EIA0384 (Washington, DC, various years).

Table 20. U.S. Methane Emissions from Mobile Sources, 1990 and 1996-2004

Source	1990	1996	1997	1998	1999	2000	2001	2002	2003	P2004
	Thousand Metric Tons Carbon Dioxide Equivalent									
Motor Vehicles										
Passenger Cars	3,284	2,505	2,464	2,233	2,212	2,180	2,134	2,091	1,913	1,903
Buses	21	24	25	26	28	28	26	25	25	25
Motorcycles	92	95	97	99	102	101	93	92	92	92
Light-Duty Trucks	1,402	1,601	1,665	1,577	1,548	1,528	1,497	1,496	1,465	1,537
Other Trucks	271	339	354	363	375	380	387	397	400	400
Total	**5,070**	**4,564**	**4,606**	**4,299**	**4,265**	**4,217**	**4,136**	**4,102**	**3,893**	**3,956**
Other Transport	525	518	494	473	494	519	489	483	434	402
Total Transport	**5,596**	**5,082**	**5,100**	**4,773**	**4,759**	**4,736**	**4,626**	**4,585**	**4,328**	**4,358**
	Thousand Metric Tons Methane									
Motor Vehicles										
Passenger Cars	143	109	107	97	96	95	93	91	83	83
Buses	1	1	1	1	1	1	1	1	1	1
Motorcycles	4	4	4	4	4	4	4	4	4	4
Light-Duty Trucks	61	70	72	69	67	66	65	65	64	67
Other Trucks	12	15	15	16	16	17	17	17	17	17
Total	**220**	**198**	**200**	**187**	**185**	**183**	**180**	**178**	**169**	**172**
Other Transport	23	23	21	21	21	23	21	21	19	17
Total Transport	**243**	**221**	**222**	**208**	**207**	**206**	**201**	**199**	**188**	**189**

P = preliminary data.

Note: Data in this table are revised from the data contained in the previous EIA report, Emissions of Greenhouse Gases in the United States 2003, DOE/EIA0573(2003) (Washington, DC, December 2004).

Sources: For passenger cars and lightduty trucks, 19902000 vehicle miles traveled (VMT) data are based on 2002 data on vehicle stocks provided by R.L. & Polk Co., with VMT modified by Oak Ridge National Laboratory (ORNL), Transportation Energy Data Book: Edition 23 (Oak Ridge, TN, October 2003), Chapter 7. 19962000 data were further adjusted using fleet data and survival curves for the population of aging vehicles. For years after 2000, emissions data are based on fleet data, econometrically modeled VMT, and survival curves for the population of aging vehicles. Calculations for buses, motorcycles, and other trucks are based on VMT from Federal Highway Administration, U.S. Department of Transportation, Federal Highway Statistics, Table VM1 (various years). Vehicle emissions coefficients are from Intergovernmental Panel on Climate Change, Greenhouse Gas Inventory Reference Manual: Revised 1996 IPCC Guidelines for National Greenhouse Gas Inventories, Vol. 3 (Paris, France, 1997), pp. 1.651.75, web site www.ipcc.ch/pub/guide.htm. Fuel consumption data for nonhighway sources are from Energy Information Administration, Fuel Oil and Kerosene Sales, DOE/EIA0535 (Washington, DC, various years); Petroleum Supply Annual, DOE/EIA0340 (Washington, DC, various years), and ORNL, Transportation Energy Data Book: Edition 23 (Oak Ridge, TN, October 2003), Chapter 9, web site http://wwwcta.ornl. gov/data/chapter9.html.

Table 21. U.S. Methane Emissions from Landfills, 1990 and 1996-2004

Type	1990	1996	1997	1998	1999	2000	2001	2002	2003	P2004
	Million Metric Tons Carbon Dioxide Equivalent									
Gross Emissions from MSW Landfills	260.5	258.3	257.5	257.0	256.4	261.4	265.2	270.8	277.3	284.8
Emissions from Industrial Landfills	18.2	18.1	18.0	18.0	17.9	18.3	18.6	19.0	19.4	19.9
Methane Recovered for Energy (-)	15.6	26.2	31.8	38.0	42.1	46.9	52.7	55.7	59.8	63.3
Methane Assumed Flared (-)	6.1	30.3	35.7	41.9	44.9	49.5	55.3	60.6	58.9	58.9
Net Emissions	**257.0**	**219.8**	**208.1**	**195.0**	**187.4**	**183.3**	**175.8**	**173.4**	**178.1**	**182.6**
	Million Metric Tons Methane									
Gross Emissions from MSW Landfills	11.3	11.2	11.2	11.2	11.1	11.4	11.5	11.8	12.1	12.4
Emissions from Industrial Landfills	0.8	0.8	0.8	0.8	0.8	0.8	0.8	0.8	0.8	0.9
Methane Recovered for Energy (-)	0.7	1.1	1.4	1.7	1.8	2.0	2.3	2.4	2.6	2.8
Methane Assumed Flared (-)	0.3	1.3	1.6	1.8	2.0	2.2	2.4	2.6	2.6	2.6
Net Emissions	**11.2**	**9.6**	**9.0**	**8.5**	**8.1**	**8.0**	**7.6**	**7.5**	**7.7**	**7.9**

*Less than 50,000 metric tons carbon dioxide equivalent.

P = preliminary data.

Note: Data in this table are revised from the data contained in the previous EIA report, *Emissions of Greenhouse Gases in the United States 2003,* DOE/EIA0573(2003) (Washington, DC, December 2004).

Sources: Municipal solid waste landfilled in 2002 from "Nationwide Survey: The State of Garbage in America," *Biocycle* (January 2004). Municipal solid waste generated and landfilled in previous years from "Nationwide Survey: The State of Garbage in America," *Biocycle* (various years), adjusted on the basis of residential demolitions, to reflect exclusion of construction and demolition waste as in the 2002 data. Municipal waste landfilled in 2003 based on 2002 estimate, scaled to annual economic growth. Emissions calculations based on S.A. Thorneloe et al ., "Estimate of Methane Emissions from U.S. Landfills," Prepared for the U.S. Environmental Protection Agency, Office of Research and Development (April 1994), and D. Augenstein, "The Greenhouse Effect and U.S. Landfill Methane," *Global Environment Change* (December 1992), pp. 311328. Methane recovered and flared from U.S. Environmental Protection Agency, Landfill Methane Outreach Program, web site www.epa.gov/lmop/. Emissions from industrial landfills estimated at 7 percent of methane emissions from municipal solid waste landfills, based on U.S. Environmental Protection Agency, *In ventory of U.S. Greenhouse Gas Emissions and Sinks 19902003,* EPA430R05003 (Washington, DC, April 2005), web site http://yosemite.epa.gov/oar/globalwarming.nsf/content/ResourceCenterPublicationsGHGEmissionsUSEmissionsInventory2005.html

Table 22. U.S. Methane Emissions from Enteric Fermentation in Domesticated Animals, 1990 and 1996-2004

Animal Type	1990	1996	1997	1998	1999	2000	2001	2002	2003	P2004
	Million Metric Tons Carbon Dioxide Equivalent									
Cattle	113.6	116.4	113.5	111.5	111.8	110.9	109.7	110.2	110.8	109.9
Swine	1.6	1.7	1.8	1.9	1.9	1.8	1.8	1.9	1.9	1.9
Sheep	2.1	1.6	1.5	1.4	1.3	1.3	1.3	1.2	1.2	1.1
Goats	0.2	0.2	0.2	0.2	0.2	0.1	0.2	0.1	0.1	0.1
Horses	2.1	2.1	2.1	2.2	2.1	2.2	2.2	2.2	2.2	2.2
Total	**119.6**	**122.0**	**119.1**	**117.2**	**117.3**	**116.3**	**115.1**	**115.7**	**116.2**	**115.2**
	Million Metric Tons Methane									
Cattle	4.94	5.06	4.94	4.85	4.86	4.82	4.77	4.79	4.82	4.78
Swine	0.07	0.07	0.08	0.08	0.08	0.08	0.08	0.08	0.08	0.08
Sheep	0.09	0.07	0.06	0.06	0.06	0.06	0.06	0.05	0.05	0.05
Goats	0.01	0.01	0.01	0.01	0.01	0.01	0.01	0.01	0.01	0.01
Horses	0.09	0.09	0.09	0.09	0.09	0.09	0.10	0.10	0.10	0.10
Total	**5.20**	**5.30**	**5.18**	**5.10**	**5.10**	**5.06**	**5.01**	**5.03**	**5.05**	**5.01**

P = preliminary data.

Notes: Data in this table are revised from the data contained in the previous EIA report, *Emissions of Greenhouse Gases in the United States 2003,* DOE/EIA0573(2003) (Washington, DC, December 2004). Totals may not equal sum of components due to independent rounding.

Sources: Cattle, sheep, and pig population data provided by the U.S. Department of Agriculture, National Agricultural Statistics Service, Livestock, Dairy and Poultry Service. Goat and horse population figures extrapolated from U.S. Department of Commerce, Bureau of the Census, *Census of Agriculture,* 1982, 1987, 1992, and 1997. Emissions calculations based on U.S. Environmental Protection Agency, *Inventory of U.S. Greenhouse Gas Emissions and Sinks 19902003,* EPA430R05003 (Washington, DC, April 2005), web site http://yosemite.epa.gov/oar/globalwarming. nsf/content/ResourceCenterPublications GHGEmissions USEmissionsInventory2005.html; and P.J. Crutzen, I. Aselmann, and W.S. Seiler, "Methane Production by Domestic Animals, Wild Ruminants, Other Herbivorous Fauna, and Humans," *Tellus,* Vol. 38B (1986), pp. 271 284.

Table 23. U.S. Methane Emissions from the Solid Waste of Domesticated Animals, 1990 and 1996-2004

Animal Type	1990	1996	1997	1998	1999	2000	2001	2002	2003	P2004
	Thousand Metric Tons Carbon Dioxide Equivalent									
Cattle										
Beef Cattle	4,037	4,814	4,682	4,592	4,547	4,496	4,470	4,454	4,433	4,406
Dairy Cattle	12,717	15,992	16,829	16,970	17,092	17,430	17,746	18,014	18,185	18,267
Swine										
Market Swine	19,222	20,543	22,522	23,366	22,522	22,368	22,627	22,724	23,165	23,522
Breeding Swine	3,502	3,681	3,893	3,739	3,488	3,507	3,470	3,390	3,362	3,340
Poultry										
Layers	1,663	1,838	1,871	1,913	1,979	2,016	2,052	2,076	2,082	2,114
Broilers	1,590	2,199	2,266	2,298	2,372	2,255	2,284	2,339	2,312	2,379
Other Animals										
Sheep	115	33	31	31	28	28	27	26	25	24
Goats	16	16	14	12	11	11	12	10	10	10
Horses	624	633	636	644	636	645	652	652	652	652
Total	**43,486**	**49,750**	**52,744**	**53,564**	**52,676**	**52,756**	**53,340**	**53,685**	**54,226**	**54,715**
	Thousand Metric Tons Methane									
Cattle										
Beef Cattle	176	209	204	200	198	195	194	194	193	192
Dairy Cattle	553	695	732	738	743	758	772	783	791	794
Swine										
Market Swine	836	893	979	1,016	979	973	984	988	1,007	1,023
Breeding Swine	152	160	169	163	152	152	151	147	146	145
Poultry										
Layers	72	80	81	83	86	88	89	90	91	92
Broilers	69	96	99	100	103	98	99	102	101	103
Other Animals										
Sheep	5	1	1	1	1	1	1	1	1	1
Goats	1	1	1	1	*	*	1	*	*	*
Horses	27	28	28	28	28	28	28	28	28	28
Total	**1,891**	**2,163**	**2,293**	**2,329**	**2,290**	**2,294**	**2,319**	**2,334**	**2,358**	**2,379**

*Less than 500 metric tons methane.

P = preliminary data.

Notes: Data in this table are revised from the data contained in the previous EIA report, *Emissions of Greenhouse Gases in the United States 2003,* DOE/EIA0573(2003) (Washington, DC, December 2004). Totals may not equal sum of components due to independent rounding.

Sources: Population data for horses and goats extrapolated from U.S. Department of Commerce, Bureau of the Census, *Census ofAgriculture,* 1982, 1987, 1992, and 1997. Population data for all other animals from U.S. Department of Agriculture, National Agricultural Statistics Service, Livestock, Dairy and Poultry Branch. Typical animal sizes from U.S. Environmental Protection Agency, Office of Air and Radiation, *Anthropogenic Methane Emissions in the United States: Estimates for 1990, Report to Congress* (Washington, DC, April 1993), p. 68; and U.S. Environmental Protection Agency, *Inventory of U.S. Greenhouse Gas Emissions and Sinks 19902003,* EPA430R05003 (Washington, DC, April2005), Table M2, web site http://yosemite.epa.gov/oar/globalwarming.nsf/content/ResourceCenterPublicationsGHGEmissionsUSEmissionsInventory2005.html. Cattle sizes adjusted by annual slaughter weight from U.S. Department of Agriculture, National Agricultural Statistics Service, Livestock, Dairy and Poultry Branch. Maximum methane production, and waste management systems used from L.M. Safley, M.E. Casada, et al., *Global Methane Emissions from Livestock and Poultry Manure* (Washington, DC: U.S. Environmental Protection Agency, February 1992), pp. 2427; U.S. Environmental Protection Agency, *Cost Methodology Report for Beef and Dairy Animal Feeding Operations,* EPA821R01019 (Washington, DC, January 2001), pp.1 13–114; and U.S. Environmental Protection Agency, *Inventory of U.S. Greenhouse Gas Emissions and Sinks 19902002,* EPA430R04003 (Washington, DC, April 2004), Table M2. General methane conversion factors from Intergovernmental Panel on Climate Change, *Greenhouse Gas Inventory Reference Manual: Revised 1996 IPCC Guidelines for National Greenhouse Gas Inventories,* Vol.3 (Paris, France, 1997), p.4.25, web site www.ipcc.ch/pub/guide.htm. State methane conversion factors for dairy cattle from U.S. Environmental Protection Agency, *In ventory of U.S.*

Greenhouse Gas Emissions and Sinks 19901998, EPA236R00001 (Washington, DC, April 2001); and U.S. Environmental Protection Agency, *Inventory of U.S. Greenhouse Gas Emissions and Sinks 19902002,* EPA430R04003 (Washington, DC, April 2004), Table M4.

Table 24. U.S. Methane Emissions from Industrial Processes, 1990 and 1996-2004

Source	1990	1996	1997	1998	1999	2000	2001	2002	2003	P2004
	Thousand Metric Tons Carbon Dioxide Equivalent									
Chemical Production										
Ethylene	380	511	531	540	578	521	470	493	479	535
Ethylene Dichloride	58	79	84	82	91	82	78	78	83	101
Styrene	335	496	476	477	499	452	357	415	437	456
Methanol	174	245	267	262	254	203	142	137	132	123
Carbon Black	331	395	402	407	415	384	363	386	386	386
Total	**1,277**	**1,726**	**1,760**	**1,767**	**1,837**	**1,643**	**1,410**	**1,509**	**1,518**	**1,602**
Iron and Steel Production										
Coke[a]	251	193	172	163	148	155	130	132	132	132
Sinter	141	136	132	125	127	124	106	104	103	93
Pig Iron	1,028	1,023	1,027	998	958	991	872	833	841	875
Total	**1,420**	**1,352**	**1,330**	**1,286**	**1,233**	**1,271**	**1,108**	**1,068**	**1,077**	**1,100**
Total Industrial Processes	**2,697**	**3,078**	**3,090**	**3,053**	**3,070**	**2,914**	**2,518**	**2,577**	**2,594**	**2,702**
	Thousand Metric Tons Methane									
Chemical Production										
Ethylene	17	22	23	23	25	23	20	21	21	23
Ethylene Dichloride	3	3	4	4	4	4	3	3	4	4
Styrene	15	22	21	21	22	20	16	18	19	20
Methanol	8	11	12	11	11	9	6	6	6	5
Carbon Black	14	17	17	18	18	17	16	17	17	17
Total	**56**	**75**	**77**	**77**	**80**	**71**	**61**	**66**	**66**	**70**
Iron and Steel Production										
Coke[a]	11	8	7	7	6	7	6	6	6	6
Sinter	6	6	6	5	6	5	5	5	4	4
Pig Iron	45	44	45	43	42	43	38	36	37	38
Total	**62**	**59**	**58**	**56**	**54**	**55**	**48**	**46**	**47**	**48**
Total Industrial Processes	**117**	**134**	**134**	**133**	**133**	**127**	**109**	**112**	**113**	**117**

[a]Based on total U.S. production of metallurgical coke, including noniron and steel uses.

P = preliminary data.

Notes: Data in this table are revised from the data contained in the previous EIA report, *Emissions of Greenhouse Gases in the United States 2003,* DOE/EIA0573(2003) (Washington, DC, December 2004). Totals may not equal sum of components due to independent rounding.

Sources: American Iron and Steel Institute, *Annual Statistical Report* (Washington, DC, various years); American Chemical Council (formerly the Chemical Manufacturers Association), *U.S. Chemical Industry Statistical Handbook* (Washington, DC, various years); and Intergovernmental Panel on Climate Change, *Greenhouse Gas Inventory Reference Manual: Revised 1996 IPCC Guidelines for National Greenhouse Gas Inventories,* Vol.3 (Paris, France, 1997), p. 2.23, web site www.ipcc/pub/guide.htm.

Nitrous Oxide Emissions

Overview

U.S. Anthropogenic Nitrous Oxide Emissions, 1990-2004

	Nitrous Oxide	Carbon Dioxide Equivalent
Estimated 2004 Emissions (Thousand Metric Tons)	1,195	353,675
Change Compared to 2003 (Thousand Metric Tons)	62	18,452
Change from 2003 *(Percent)*	*5.5%*	*5.5%*
Change Compared to 1990 (Thousand Metric Tons)	56	16,638
Change from 1990 *(Percent)*	*4.9%*	*4.9%*

Estimated U.S. anthropogenic nitrous oxide emissions totaled 1.2 million metric tons in 2004, or 353.7 million metric tons carbon dioxide equivalent ($MMTCO_2e$), 5.5 percent more than in 2003 and 4.9 percent above 1990 levels (Table 25). The 2004 total for nitrous oxide emissions represents 5.0 percent of all U.S. greenhouse gas emissions for the year. Most of the increase in U.S. nitrous oxide emissions for 2004 can be attributed to emissions from agricultural sources, which increased by 17.4 $MMTCO_2e$ of nitrous oxide or 94 percent of the overall increase in nitrous oxide emissions of 18.5 $MMTCO_2e$.

Sources **of U.S. Anthropogenic Nitrous Oxide Emissions, 1990-2004**

	Million Metric Tons CO_2e		*Percent Change*	
Source	**1990**	**2004**	**1990-2004**	**2003-2004**
Agriculture	249.3	265.2	*6.4%*	*7.0%*
Energy	54.4	68.4	*25.9%*	*1.5%*
Industrial Processes	28.6	14.0	*-51.1%*	*-0.3%*
Waste Management	4.8	6.0	*24.8%*	*1.1%*

A downward trend in nitrous oxide emissions that began in 1995, after emissions of nitrous oxide peaked at 374.5 $MMTCO_2e$ in 1994, was ended in 2003. With the increase in 2004, annual U.S. emissions of nitrous oxide were higher than their 1990 level (337.0 $MMTCO_2e$) for the first time since 2000.

Sources of U.S. nitrous oxide emissions include energy use, agriculture, waste management, and industrial processes. The largest component of U.S. anthropogenic nitrous oxide emissions is emissions from agricultural activities at 265.2 $MMTCO_2e$ or 75 percent of total nitrous oxide emissions. Nitrogen fertilization of agricultural soils represents 77 percent of emissions from agricultural activities. Most of the remainder (23 percent) is from the handling of animal waste in managed systems. Small quantities of nitrous oxide (0.2 percent of nitrous oxide emissions from agricultural activities) are also released from the burning of crop residues. Estimated emissions of nitrous oxide from agricultural sources in 2004 were 7.0 percent above 2003 levels and 6.4 percent above 1990 levels (Figure 3).

The secondlargest source of anthropogenic nitrous oxide emissions is energy consumption, which includes mobile source combustion from passenger cars, buses, motorcycles, and trucks; and stationary source combustion from commercial, residential, industrial, and electric power sector energy use. Energy use was responsible for 68.4 $MMTCO_2e$ of nitrous oxide emissions in 2004 (19 percent of total U.S. nitrous oxide emissions).

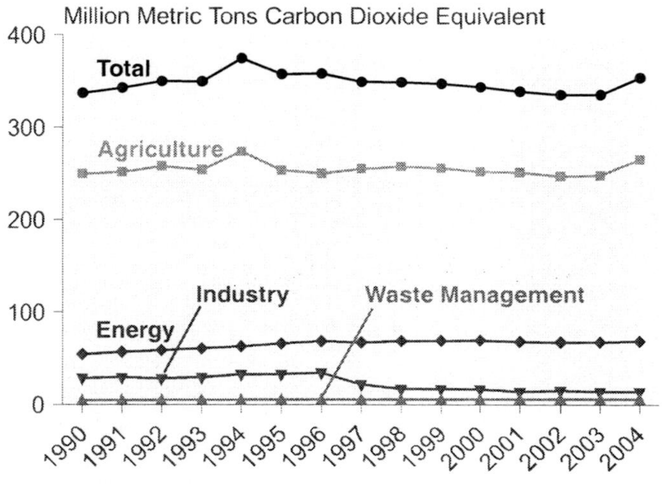

Source: Estimates presented in this chapter.

Figure 3. U.S. Emissions of Nitrous Oxide by Source, 1990-2004

Although the 2004 level of emissions from energy sources is 1.5 percent higher than the 2003 level, it is 26 percent higher than in 1990.

Industrial production of adipic and nitric acid, which releases nitrous oxide as a byproduct, accounted for emissions of 14.0 $MMTCO_2e$ in 2004 (3.9 percent of total U.S. nitrous oxide emissions), slightly (0.3 percent) lower than the 2003 level and 51 percent lower than in 1990. The large decline in emissions of nitrous oxide from adipic acid production since 1990 is a result of the continuing utilization of emissions control technology at three of the four adipic acid plants operating in the United States.

Nitrous oxide emissions from activities related to waste management in 2004 totaled 6.0 $MMTCO_2e$, or 1.7 percent of all U.S. anthropogenic nitrous oxide emissions (Table 25). During 2004, emissions from human sewage in wastewater accounted for 96 percent of estimated nitrous oxide emissions from this source, and the remainder was associated with waste combustion.

Agriculture

U.S. Nitrous Oxide Emissions from Agriculture, 1990-2004

Estimated 2004 Emissions (Million Metric Tons Carbon Dioxide Equivalent)	265.2
Change Compared to 2003 (Million Metric Tons Carbon Dioxide Equivalent)	17.4
Change from 2003 *(Percent)*	*7.0%*
Change Compared to 1990 (Million Metric Tons Carbon Dioxide Equivalent)	16.0
Change from 1990 *(Percent)*	*6.4%*

Nitrous oxide emissions from agricultural activities increased by 17.4 $MMTCO_2e$ (7.0 percent) in 2004 to a total of 265.2 $MMTCO_2e$, compared with 247.8 $MMTCO_2e$ in 2003. Since 1990, nitrous oxide emissions from agricultural activities have increased by 6.4 percent. Agricultural activities were responsible for 75 percent of U.S. nitrous oxide emissions in 2004, smaller than the 86percent share that agricultural practices contribute to nitrous oxide emissions globally.[79] Nitrous oxide emissions from agricultural activities represent 3.7 percent of total U.S. greenhouse gas emissions.

Nitrogen fertilization of agricultural soils accounted for 77 percent of U.S. agricultural emissions of nitrous oxide in 2004. Nearly all the remaining agricultural emissions (23 percent) can be traced to the management of the solid waste of domesticated animals. The disposal of crop residues by burning also produces nitrous oxide that is released into the atmosphere; however, the amount is relatively minor, at 0.6 $MMTCO_2e$ (2 thousand metric tons nitrous oxide) or 0.2 percent of total U.S. emissions of nitrous oxide from agricultural sources in 2004.

Nitrogen Fertilization of Agricultural Soils

EIA estimates that 204.3 $MMTCO_2e$ of nitrous oxide was released into the atmosphere as a result of direct and indirect emissions associated with fertilization practices in 2004 (Table 26). Estimated emissions increased by 9.5 percent compared with 2003 levels and were 9.3 percent higher than in 1990.

Nitrogen uptake and nitrous oxide emissions occur naturally as a result of nitrification and denitrification processes in soil and crops, generally through bacterial action. When nitrogen compounds are added to the soil, bacterial action is stimulated, and emissions generally increase, unless the application precisely matches plant uptake and soil capture.[80] Nitrogen may be added to the soil by synthetic or organic fertilizers, nitrogenfixing crops, and crop residues. Nitrogenrich soils, called "histosols," may also stimulate emissions. Direct emissions in 2004 (163.9 $MMTCO_2e$) represented 80 percent of total emissions from nitrogen fertilization, with the primary components including the biological fixation of nitrogen in crops (70.9 $MMTCO_2e$), nitrogen fertilizers (50.0 $MMTCO_2e$), and crop residues (38.5 $MMTCO_2e$).

Indirect emissions from nitrogen fertilization result from adding excess nitrogen to the soil, which in turn enriches ground and surface waters, such as rivers and streams, and results in emissions of nitrous oxide. This source is referred to as "soil leaching." Additional indirect emissions occur from "atmospheric deposition," in which soils emit other nitrogen compounds that react to form nitrous oxide in the atmosphere. Indirect emissions in 2004 (40.4 $MMTCO_2e$) represented 20 percent of total emissions from nitrogen fertilization, with soil leaching accounting for 34.3 $MMTCO_2e$ and atmospheric deposition totaling 6.1 $MMTCO_2e$.

There are significant uncertainties associated with estimating the amount of emissions produced by adding nitrogen to agricultural soils. Models used to estimate the amount are based on limited sources of experimental data.[81] The uncertainty increases when moving from emissions associated with animal manure to soil mineralization and atmospheric deposition, where both estimating the amount of emissions and segmenting anthropogenic from biogenic sources become increasingly difficult.

Solid Waste of Domesticated Animals

Estimated 2004 nitrous oxide emissions from animal waste management were 60.3 $MMTCO_2e$, down by 0.6 percent from 2003 levels and 2.6 percent lower than 1990 levels (Table 27), making animal waste management the secondlargest U.S. agricultural source of nitrous oxide emissions, after nitrogen fertilization of soils. Nitrous oxide emissions from animal waste are dominated by emissions from cattle waste, which in 2004 accounted for 93 percent of emissions from the solid waste of domesticated animals (a total of 55.8 $MMTCO_2e$ in 2004).

Nitrous oxide is released as part of the microbial denitrification of animal manure. The total volume of nitrous oxide emissions is a function of animal size and manure production, the amount of nitrogen in the animal waste, and the method of managing the animal waste. Waste managed by a solid storage or pasture range method may emit 20 times more nitrous oxide per unit of nitrogen content than does waste managed in anaerobic lagoon and liquid systems. Generally, solid waste from feedlot beef cattle is managed with the solid storage or pasture range method, accounting for the majority of nitrous oxide emissions. Solid waste from swine is generally managed in anaerobic lagoons and other liquid systems. Anaerobic digestion yields methane emissions but only negligible amounts of nitrous oxide. Thus, changes in estimated emissions result primarily from changes in cattle populations. For example, beef cattle populations grew during the first half of the 1990s, leading to higher emissions through 1995, but have since declined slowly, lowering emissions to below 1990 levels.

Crop Residue Burning

In 2004, estimated emissions of nitrous oxide from crop residue burning were 0.6 $MMTCO_2e$, up by 0.1 $MMTCO_2e$ (17 percent) from 2003 levels (Table 25). The large percentage increase is mainly attributable to increased corn and soybean production. Emissions from this source remain very small, at 0.2 percent of all U.S. nitrous oxide emissions. When crop residues are burned, the incomplete combustion of agricultural waste results in the production of nitrous oxide, as well as methane (discussed in Chapter 3).

Energy Use

U.S. Nitrous Oxide Emissions from Energy, 1990-2004

Estimated 2004 Emissions (Million Metric Tons Carbon Dioxide Equivalent)	68.4
Change Compared to 2003 (Million Metric Tons Carbon Dioxide Equivalent)	1.0
Change from 2003 *(Percent)*	*1.5%*
Change Compared to 1990 (Million Metric Tons Carbon Dioxide Equivalent)	14.1
Change from 1990 *(Percent)*	*25.9%*

The energy use category includes nitrous oxide emissions from both mobile and stationary sources as byproducts of fuel combustion. Estimated 2004 energyrelated emissions were 68.4 MMTCO_2e, or 19 percent of total U.S. anthropogenic nitrous oxide emissions (Table 25). Emissions from energy use are dominated by mobile combustion (79 percent of nitrous oxide emissions from energy use in 2004).

Mobile Combustion

Nitrous oxide emissions from mobile source combustion in 2004 were 53.8 MMTCO_2e, an increase of 1.5 percent from the 2003 level of 53.0 MMTCO_2e (Table 28). In addition to emissions from passenger cars and lightduty trucks, emissions from air, rail, and marine transportation and from farm and construction equipment are also included in the estimates. Motor vehicles, however, are the predominant source, accounting for 94 percent of nitrous oxide emissions from mobile combustion (Table 27).

Nitrous oxide emissions from motor vehicles are caused primarily by the conversion of nitrogen oxides (NO_x) into nitrous oxide (N_2O) by vehicle catalytic converters. The normal operating temperature of catalytic converters is high enough to cause the thermal decomposition of nitrous oxide. Consequently, it is probable that nitrous oxide emissions result primarily from "cold starts" of motor vehicles and from catalytic converters that are defective or operating under abnormal conditions. This implies that the primary determinant of the level of emissions is motor vehicle operating conditions; however, different types of catalytic converters appear to differ systematically in their emissions, and emissions probably vary with engine size. Thus, emissions also depend on the "mix" of vehicle age and type on the road.

Nitrous oxide emissions from mobile sources grew rapidly between 1990 and 1996 due to increasing motor vehicle use, the shifting composition of the lightduty vehicle fleet toward light trucks that have lower fuel economy and higher permile emission factors, and the gradual replacement of low emitting pre1983 vehicles that did not use catalytic converters with higher emitting post1983 vehicles that do use catalytic converters. This growth moderated between 1996 and 1999 due to the introduction of more advanced, loweremitting catalytic converters. After peaking in 1999, emissions declined slowly through 2002, as vehicle turnover led to a fleet dominated by the more advanced catalytic converters.

Stationary Combustion

In 2004, estimated nitrous oxide emissions from stationary combustion sources were 14.7 $MMTCO_2e$, 1.5 percent (0.2 $MMTCO_2e$) higher than in 2003 and 10 percent (1.4 $MMTCO_2e$) higher than in 1990 (Table 29). The total emissions increase from this source between 1990 and 2004 (1.4 $MMTCO_2e$) can be attributed principally to coalfired combustion systems. Nitrous oxide emissions from coalfired combustion systems increased by 17 percent over the period, from 8.0 $MMTCO_2e$ in 1990 to 9.3 $MMTCO_2e$ in 2004.

Coalfired combustion systems produced 64 percent (9.3 $MMTCO_2e$) of the 2004 emissions of nitrous oxide from stationary combustion. Other fuels—including fuel oil (2.4 $MMTCO_2e$), wood (2.3 $MMTCO_2e$), and natural gas (0.6 $MMTCO_2e$)—accounted for the balance. During combustion, nitrous oxide is produced as a result of chemical interactions between nitrogen oxides (mostly NO_2) and other combustion products. With most conventional stationary combustion systems, high temperatures destroy almost all nitrous oxide, limiting the quantity that escapes; therefore, emissions from these systems typically are low.

Industrial Sources

Emissions of nitrous oxide from industrial sources were 14.0 $MMTCO_2e$ in 2004, a decrease of less than 0.05 $MMTCO_2e$ (0.3 percent) from 2003 and a decrease of 14.6 $MMTCO_2e$ (51 percent) since 1990. Nitrous oxide is emitted as a byproduct of certain chemical production processes. Table 30 provides estimates of emissions from the production of adipic acid and nitric acid, the two principal known sources.

Nitric Acid Production

The 6.7 million metric tons of nitric acid manufactured in 2004[82] resulted in estimated nitrous oxide emissions of 10.9 $MMTCO_2e$ (Table 30). This estimate was 0.7 percent lower than 2003 levels and 6.8 percent lower than 1990 levels. The emissions factor used to estimate nitrous oxide emissions from the production of nitric acid was based on measurements at a single DuPont plant, which indicated an emissions factor of 2 to 9 grams of nitrous oxide emitted per kilogram of nitric acid manufactured, suggesting a significant range of uncertainty. Nitric acid, a primary ingredient in fertilizers, usually is manufactured by oxidizing ammonia (NH_3) with a platinum catalyst. Nitrous oxide emissions are a direct result of the oxidation.

Adipic Acid Production

Emissions from adipic acid production in 2004 were 3.0 $MMTCO_2e$, 1 percent higher than in 2003. Nitrous oxide emissions from this source in 2004 were 82 percent (13.8 $MMTCO_2e$) lower than in 1990.

U.S. Nitrous Oxide Emissions from Industrial Sources, 1990-2004

Estimated 2004 Emissions (Million Metric Tons Carbon Dioxide Equivalent)	14.0
Change Compared to 2003 (Million Metric Tons Carbon Dioxide Equivalent)	*
Change from 2003 *(Percent)*	*-0.3%*
Change Compared to 1990 (Million Metric Tons Carbon Dioxide Equivalent)	-14.6
Change from 1990 *(Percent)*	*-51.1%*

*Less than 0.05 million metric tons.

Adipic acid is a fine white powder that is used primarily in the manufacture of nylon fibers and plastics, such as carpet yarn, clothing, and tire cord. Other uses of adipic acid include production of plasticizer for polyvinyl chloride and polyurethane resins, lubricants, insecticides, and dyes. In the United States, three companies, which operate four plants, manufacture adipic acid by oxidizing a ketonealcohol mixture with nitric acid. Nitrous oxide is an intrinsic byproduct of this chemical reaction. For every metric ton of adipic acid produced, 0.3 metric ton of nitrous oxide is created.[83] Between 1990 and 1996, emissions from adipic acid manufacture grew by 23 percent, reaching 20.7 $MMTCO_2e$ before dropping sharply to 7.8 $MMTCO_2e$ in 1997 (Table 30).

Before 1997, two of the four plants that manufacture adipic acid controlled emissions by thermally decomposing the nitrous oxide. This technique eliminates 98 percent of potential nitrous oxide emissions from the process.[84] During the first quarter of 1997, a third plant installed emissions controls, increasing the share of adipic acid production employing emissions abatement controls from 74 percent in 1996 to 92 percent in 1997. In 1998, with emissions controls in place for the full year, 97 percent of emissions from U.S. adipic acid production were controlled.[85]

Estimated emissions of nitrous oxide from uncontrolled adipic acid production decreased from 19.6 $MMTCO_2e$ in 1996 to 2.0 $MMTCO_2e$ in 1998 and remained fairly stable through 2002, before dropping to 1.4 $MMTCO_2e$ in 2003 and 2004. Emissions of nitrous oxide from controlled plants have remained relatively constant from 1998 through 2004, in a range of 1.4 to 1.6 $MMTCO_2e$. With the share of adipic acid production employing abatement controls now at 98 percent, future changes in nitrous oxide emissions from this source are expected to result primarily from changes in plant production levels in response to market demand.

Waste Management

Nitrous oxide emissions from waste management in 2004 are estimated at 6.0 $MMTCO_2e$, or 1.7 percent of all U.S. anthropogenic nitrous oxide emissions (Table 25). During 2004, emissions from human sewage in wastewater were responsible for 96 percent of the estimated emissions from this source, and the remainder was associated with waste combustion.

Estimated emissions from waste management increased by 1.1 percent between 2003 and 2004 and by 25 percent between 1990 and 2004. Because of the lack of reliable data and an effective estimation method, no estimate of emissions from industrial wastewater was calculated, leaving estimated emissions from waste management lower than they otherwise would be had a viable estimation method been available.

Human Sewage in Wastewater

In 2004, nitrous oxide emissions from wastewater were 5.8 $MMTCO_2e$, a 1.0percent increase from 2003 levels and a 26percent increase from the 1990 level (Table 25). Estimates of nitrous oxide emissions from human waste are scaled to population size and per capita protein intake. U.S. population has grown by 18 percent since 1990. U.S. per capita protein intake rose steadily between 1990 and 1999, before declining between 2000 and 2003. Today, U.S. per capita protein intake is 6.5 percent above 1990 levels. Data on protein intake are taken from the United Nations Food and Agriculture Organization (FAO).[86]

Nitrous oxide is emitted from wastewater that contains nitrogenbased organic materials, such as those found in human or animal waste. Two natural processes—nitrification and denitrification—combine to produce nitrous oxide. Nitrification, an aerobic process, converts ammonia into nitrate; denitrification, an anaerobic process, converts nitrate to nitrous oxide. Factors that influence the amount of nitrous oxide generated from wastewater include temperature, acidity, biochemical oxygen demand (BOD),[87] and nitrogen concentration.

Waste Combustion

In 2004, estimated nitrous oxide emissions from waste combustion were 0.2 $MMTCO_2e$, up by 4.8 percent from the 2003 level and 6.1 percent above the 1990 level. Data on the amount of waste generated in the United States in 2004 were not available in time for this report; therefore, EIA scaled the 2004 estimate for waste combustion to the growth in U.S. gross domestic product. The share of waste burned is estimated to have been unchanged from 2003 to 2004, and the total volume of waste generated is estimated to have increased by 4.8 percent. The total volume of waste generated in the United States increased by 58 percent between 1990 and 2004; however, the share of waste burned in 2004 was just 7.7 percent, compared with 12 percent in 1990.

U.S. Nitrous Oxide Emissions from Waste Management, 1990-2004

Estimated 2004 Emissions (Million Metric Tons Carbon Dioxide Equivalent)	6.0
Change Compared to 2003 (Million Metric Tons Carbon Dioxide Equivalent)	0.1
Change from 2003 *(Percent)*	*1.1%*
Change Compared to 1990 (Million Metric Tons Carbon Dioxide Equivalent)	1.2
Change from 1990 *(Percent)*	*24.8%*

Table 25. Estimated U.S. Emissions of Nitrous Oxide, 1990 and 1996-2004

Source	1990	1996	1997	1998	1999	2000	2001	2002	2003	P2004
Million Metric Tons Carbon Dioxide Equivalent										
Agriculture										
Nitrogen Fertilization of Soils	186.9	184.2	190.5	194.0	192.7	189.4	189.1	185.1	186.6	204.3
Solid Waste of Domesticated Animals	61.9	65.1	63.8	62.9	62.4	61.8	61.4	61.2	60.7	60.3
Crop Residue Burning	0.5	0.5	0.6	0.6	0.5	0.6	0.6	0.5	0.5	0.6
Subtotal	**249.3**	**249.8**	**254.8**	**257.4**	**255.6**	**251.8**	**251.1**	**246.8**	**247.8**	**265.2**
Energy Use										
Mobile Combustion	41.1	53.9	52.4	54.2	54.3	54.3	53.3	53.0	53.0	53.8
Stationary Combustion	13.3	14.5	14.6	14.5	14.7	15.1	14.5	14.2	14.5	14.7
Subtotal	**54.4**	**68.5**	**67.0**	**68.7**	**68.9**	**69.4**	**67.9**	**67.2**	**67.5**	**68.4**
Industrial Sources	**28.6**	**34.3**	**21.8**	**17.2**	**16.8**	**16.6**	**14.0**	**15.2**	**14.0**	**14.0**
Waste Management										
Human Sewage in Wastewater	4.6	5.2	5.2	5.3	5.5	5.6	5.6	5.7	5.7	5.8
Waste Combustion	0.2	0.2	0.2	0.2	0.2	0.2	0.2	0.2	0.2	0.2
Subtotal	**4.8**	**5.4**	**5.4**	**5.5**	**5.7**	**5.8**	**5.9**	**5.9**	**6.0**	**6.0**
Total	**337.0**	**358.0**	**349.1**	**348.8**	**347.1**	**343.5**	**338.8**	**335.1**	**335.2**	**353.7**
Thousand Metric Tons Nitrous Oxide										
Agriculture										
Nitrogen Fertilization of Soils	631	622	643	655	651	640	639	625	630	690
Solid Waste of Domesticated Animals	209	220	216	212	211	209	208	207	205	204
Crop Residue Burning	2	2	2	2	2	2	2	2	2	2
Subtotal	**842**	**844**	**861**	**870**	**864**	**851**	**848**	**834**	**837**	**896**
Energy Use										
Mobile Combustion	139	182	177	183	183	183	180	179	179	182
Stationary Combustion	45	49	49	49	50	51	49	48	49	50
Subtotal	**184**	**231**	**227**	**232**	**233**	**234**	**229**	**227**	**228**	**231**
Industrial Sources	**96**	**116**	**74**	**58**	**57**	**56**	**47**	**51**	**47**	**47**
Waste Management										
Human Sewage in Wastewater	16	17	18	18	19	19	19	19	19	20
Waste Combustion	1	1	1	1	1	1	1	1	1	1
Subtotal	**16**	**18**	**18**	**19**	**19**	**20**	**20**	**20**	**20**	**20**
Total	**1,139**	**1,209**	**1,179**	**1,178**	**1,173**	**1,160**	**1,145**	**1,132**	**1,133**	**1,195**

P = preliminary data.

Notes: Data in this table are revised from the data contained in the previous EIA report, *Emissions of Greenhouse Gases in the United States 2003,* DOE/EIA0573(2003) (Washington, DC, December 2004). Totals may not equal sum of components due to independent rounding.

Sources: Estimates presented in this chapter. Emissions calculations based on Intergovernmental Panel on Climate Change, *Greenhouse Gas Inventory Reference Manual: Revised 1996 IPCC Guidelines for National Greenhouse Gas Inventories,* Vol. 3 (Paris, France, 1997), pp. 4.814.94, web site www.ipcc.ch/pub/guide.htm; and U.S. Environmental Protection Agency, *Inventory of U.S. Greenhouse Gas Emissions and Sinks: 19902003,* EPA 430R05003 (Washington, DC, April 2005), web site http://yosemite.epa.gov/oar/globalwarming.nsf/content/ResourceCenter PublicationsGHGEmissionsUSEmissionsInventory2005.html.

Table 26. U.S. Nitrous Oxide Emissions from Nitrogen Fertilization of Agricultural Soils, 1990 and 1996-2004

Source	1990	1996	1997	1998	1999	2000	2001	2002	2003	P2004
				Million Metric Tons Carbon Dioxide Equivalent						
Direct Emissions										
Biological Fixation in Crops. . . .	58.6	62.6	66.4	68.7	68.2	67.9	69.5	65.3	62.5	70.9
Nitrogen Fertilizers	53.1	47.1	47.2	47.5	47.7	45.6	44.4	45.6	48.0	50.0
Crop Residues	28.2	31.9	34.2	34.8	33.8	34.6	34.7	32.9	32.8	38.5
Soil Mineralization.	3.0	3.1	3.1	3.1	3.1	3.1	3.1	3.1	3.2	3.2
Animal Manure	1.2	1.3	1.3	1.3	1.3	1.3	1.3	1.3	1.3	1.3
Sewage Sludge.	0.3	0.3	0.3	0.3	0.3	0.4	0.4	0.5	0.5	0.6
Total Direct Emissions	**144.1**	**146.0**	**152.2**	**155.5**	**154.1**	**152.4**	**153.0**	**148.2**	**147.7**	**163.9**
Indirect Emissions										
Soil Leaching	36.3	32.4	32.4	32.6	32.8	31.3	30.6	31.4	33.0	34.3
Atmospheric Deposition	6.5	5.8	5.8	5.8	5.8	5.6	5.5	5.6	5.9	6.1
Total Indirect Emissions . . .	**42.8**	**38.2**	**38.2**	**38.5**	**38.6**	**36.9**	**36.0**	**37.0**	**38.8**	**40.4**
Total	**186.9**	**184.2**	**190.5**	**194.0**	**192.7**	**184.9**	**189.1**	**185.1**	**186.6**	**204.3**
				Thousand Metric Tons Nitrous Oxide						
Direct Emissions										
Biological Fixation in Crops. . . .	198	212	224	232	230	229	235	221	211	240
Nitrogen Fertilizers	179	159	159	161	161	154	150	154	162	169
Crop Residues	95	108	116	118	114	117	117	111	111	130
Soil Mineralization.	10	10	10	10	10	11	11	11	11	11
Animal Manure	4	4	4	4	4	4	4	4	4	4
Sewage Sludge.	1	1	1	1	1	1	1	2	2	2
Total Direct Emissions	**487**	**493**	**514**	**525**	**521**	**515**	**517**	**501**	**499**	**554**
Indirect Emissions										
Soil Leaching	123	109	110	110	111	106	103	106	111	116
Atmospheric Deposition	22	20	20	20	20	19	18	19	20	21
Total Indirect Emissions . . .	**144**	**129**	**129**	**130**	**130**	**125**	**122**	**125**	**131**	**136**
Total	**631**	**622**	**643**	**655**	**651**	**640**	**639**	**625**	**630**	**690**

P = preliminary data.

Notes: Data in this table are revised from the data contained in the previous EIA report, *Emissions of Greenhouse Gases in the United States 2003,* DOE/EIA0573(2003) (Washington, DC, December 2004). Totals may not equal sum of components due to independent rounding.

Sources: Estimates presented in this chapter. Emissions coefficients from Intergovernmental Panel on Climate Change, *Greenhouse Gas Inventory Reference Manual: Revised 1996 IPCC Guidelines for National Greenhouse Gas Inventories,* Vol. 3 (Paris, France, 1997), pp. 4.894.107, web site www.ipcc.ch/pub/guide.htm. Total nitrogen content of U.S. commercial fertilizer consumption—19881994, Tennessee Valley Authority; 19952002, Association of American Plant Food Control Officials, *Commercial Fertilizers* (Washington, DC, various years). Manure application based on cattle population data provided by the U.S. Department of Agriculture, National Agricultural Statistics Service, web sites www.usda.gov/nass/ pubs/histdata.htm and www.nass.usda.gov/ipedb/. Typical animal sizes from U.S. Environmental Protection Agency, Office of Air and Radiation, *Anthropogenic Methane Emissions in the United States: Estimates for 1990* (Washington, DC, April 1993), p. 68. Manure production and waste management systems used from L.M. Safley, M.E. Casada et al., *Global Methane Emissions From Livestockand Poultry Manure* (Washington, DC, February 1992), and U.S. Environmental Protection Agency, *Inventory of U.S. Greenhouse Gas Emissions and Sinks: 19902003,* EPA 430R05003 (Washington, DC, April 2005), web site http://yosemite.epa.gov/oar/globalwarming.nsf/content/ResourceCenterPublicationsGHGEmissions USEmissionsInventory2005.html.

Table 27. U.S. Nitrous Oxide Emissions from Solid Waste of Domesticated Animals, 1990 and 1996-2004

Source	1990	1996	1997	1998	1999	2000	2001	2002	2003	P2004
	Million Metric Tons Carbon Dioxide Equivalent									
Cattle	57.5	60.6	59.2	58.3	57.9	57.4	56.9	56.7	56.3	55.8
Swine	1.5	1.5	1.7	1.7	1.6	1.6	1.6	1.6	1.7	1.7
Poultry	0.9	1.2	1.3	1.3	1.3	1.3	1.3	1.3	1.3	1.4
Horses	0.7	0.7	0.7	0.7	0.7	0.7	0.7	0.7	0.7	0.7
Sheep	1.0	0.7	0.7	0.7	0.6	0.6	0.6	0.6	0.6	0.5
Goats	0.3	0.3	0.2	0.2	0.2	0.2	0.2	0.2	0.2	0.2
Total	**61.9**	**65.1**	**63.8**	**62.9**	**62.4**	**61.8**	**61.4**	**61.2**	**60.7**	**60.3**
	Thousand Metric Tons Nitrous Oxide									
Cattle	194	205	200	197	195	194	192	192	190	189
Swine	5	5	6	6	6	5	6	6	6	6
Poultry	3	4	4	4	4	4	4	4	4	5
Horses	2	2	2	2	2	2	2	2	2	2
Sheep	3	3	2	2	2	2	2	2	2	2
Goats	1	1	1	1	1	1	1	1	1	1
Total	**209**	**220**	**216**	**212**	**211**	**209**	**208**	**207**	**205**	**204**

P = preliminary data.

Note: Totals may not equal sum of components due to independent rounding.

Sources: Estimates presented in this chapter. Nitrogen content of waste by species, manure management systems, and emissions coefficients from Intergovernmental Panel on Climate Change, *Greenhouse Gas Inventory Reference Manual: Revised 1996 IPCC Guidelines for National Greenhouse Gas Inventories,* Vol. 3 (Paris, France, 1997), pp. 4.894.107, web site www.ipcc.ch/pub/guide.htm. Population data for horses and goats extrapolated from U.S. Department of Commerce, Bureau of the Census, *Census ofAgriculture* (1982, 1987, 1992, and 1997). All other animal populations from U.S. Department of Agriculture, National Agricultural Statistics Service, web sites www.usda.gov/nass/pubs/histdata.htm and www.nass.usda.gov/ipedb/. Typical animal sizes from U.S. Environmental Protection Agency, Office of Air and Radiation, *Anthropogenic Methane Emissions in the United States: Estimates for 1990* (Washington, DC, April 1993), p. 68. Cattle sizes adjusted by annual slaughter weight from U.S. Department of Agriculture, National Agricultural Statistics Service.

Table 28. U.S. Nitrous Oxide Emissions from Mobile Combustion, 1990 and 1996-2004

Item	1990	1996	1997	1998	1999	2000	2001	2002	2003	P2004
	Million Metric Tons Carbon Dioxide Equivalent									
Motor Vehicles										
Passenger Cars	25.2	30.4	29.1	30.4	30.3	30.0	29.4	28.7	28.6	28.5
Light-Duty Trucks	10.4	17.5	17.3	17.7	17.7	17.8	17.7	17.9	18.3	19.2
Other Trucks	1.7	2.2	2.3	2.3	2.4	2.4	2.5	2.6	2.6	2.6
Buses	0.1	0.1	0.1	0.1	0.1	0.1	0.1	0.1	0.1	0.1
Motorcycles	*	*	*	*	*	*	*	*	*	*
Subtotal	**37.4**	**50.2**	**48.8**	**50.5**	**50.5**	**50.4**	**49.6**	**49.3**	**49.5**	**50.4**
Other Mobile Sources	**3.6**	**3.7**	**3.6**	**3.6**	**3.7**	**3.9**	**3.7**	**3.7**	**3.5**	**3.4**
Total	**41.1**	**53.9**	**52.4**	**54.2**	**54.3**	**54.3**	**53.3**	**53.0**	**53.0**	**53.8**
	Thousand Metric Tons Nitrous Oxide									
Motor Vehicles										
Passenger Cars	85	103	98	103	102	101	99	97	97	96
Light-Duty Trucks	35	59	59	60	60	60	60	61	62	65
Other Trucks	6	7	8	8	8	8	8	9	9	9
Buses	*	*	*	*	*	*	*	*	*	*
Motorcycles	*	*	*	*	*	*	*	*	*	*
Subtotal	**127**	**170**	**165**	**171**	**171**	**170**	**168**	**167**	**167**	**170**
Other Mobile Sources	**12**	**13**	**12**	**12**	**13**	**13**	**13**	**12**	**12**	**11**
Total	**139**	**182**	**177**	**183**	**183**	**183**	**180**	**179**	**179**	**182**

*Less than 500 metric tons nitrous oxide or less than 50,000 metric tons carbon dioxide equivalent.

P = preliminary data.

Notes: Data in this table are revised from the data contained in the previous EIA report, *Emissions of Greenhouse Gases in the United States 2003,* DOE/EIA0573(2003) (Washington, DC, December 2004). Totals may not equal sum of components due to independent rounding.

Sources: Calculations based on vehicle miles traveled from U.S. Department of Transportation, *Federal Highway Statistics* (various years), Table VM1, and current year preliminary estimates calculated using growth rates from EIA, *ShortTerm Energy Outlook* (various years). Other Mobile Sources calculations based on Oak Ridge National Laboratory, *Transportation Energy Data Book;* EIA, *Fuel Oil and Kerosene Sales, State Energy Data Report,* and *Petroleum SupplyAnnual* (various years). Passenger car and lightduty truck emissions coefficients from U.S. Environmental Protection Agency, Office of Air and Radiation, *Emissions of Nitrous Oxide From Highway Mobile Sources: Comments on the Draft Inventory of U.S. Greenhouse Gas Emissions and Sinks, 19901996,* EPA420R98009 (Washington DC, August 1998). Emissions coefficients from Intergovernmental Panel on Climate Change, *Greenhouse Gas Inventory Reference Manual: Revised 1996 IPCC Guidelines for National Greenhouse Gas Inventories,* Vol. 3 (Paris, France, 1997), pp. 1.641.68, web site www.ipcc.ch/pub/guide.htm.

Table 29. U.S. Nitrous Oxide Emissions from Stationary Combustion, 1990 and 1996-2004

Source	1990	1996	1997	1998	1999	2000	2001	2002	2003	P2004
		Thousand Metric Tons Carbon Dioxide Equivalent								
Residential										
Coal	13	7	7	5	6	4	5	5	4	5
Fuel Oil[a]	251	266	255	235	263	279	275	261	268	280
Natural Gas	128	153	146	133	138	145	139	143	149	143
Wood	677	693	504	451	482	504	431	365	418	387
Total	**1,069**	**1,119**	**912**	**823**	**889**	**933**	**850**	**773**	**840**	**815**
Commercial										
Coal	52	50	54	42	42	36	38	38	35	36
Fuel Oil[a]	170	134	126	118	118	134	135	122	134	144
Natural Gas	77	92	94	88	89	92	89	92	94	88
Wood	45	58	57	56	61	62	47	45	47	48
Total	**344**	**335**	**330**	**304**	**310**	**324**	**308**	**297**	**310**	**315**
Industrial										
Coal	1,131	995	977	925	905	918	913	841	851	844
Fuel Oil[a]	1,606	1,752	1,789	1,781	1,845	1,768	1,807	1,644	1,655	1,752
Natural Gas	242	283	283	279	268	271	248	252	242	250
Wood	1,680	1,962	2,017	1,867	1,887	1,906	1,681	1,626	1,588	1,687
Total	**4,659**	**4,992**	**5,066**	**4,853**	**4,905**	**4,863**	**4,649**	**4,363**	**4,335**	**4,532**
Electric Power										
Coal	6,770	7,680	7,878	8,008	8,034	8,426	8,205	8,244	8,411	8,446
Fuel Oil[a]	228	146	166	233	216	204	229	216	215	213
Natural Gas	94	110	118	133	140	151	153	164	150	156
Wood	150	161	160	160	161	156	147	175	195	196
Total	**7,242**	**8,097**	**8,321**	**8,534**	**8,551**	**8,937**	**8,733**	**8,799**	**8,971**	**9,011**
Total All Sectors										
Coal	**7,965**	**8,732**	**8,916**	**8,980**	**8,987**	**9,384**	**9,161**	**9,128**	**9,301**	**9,331**
Fuel Oil[a]	**2,256**	**2,298**	**2,335**	**2,367**	**2,443**	**2,385**	**2,445**	**2,242**	**2,272**	**2,389**
Natural Gas	**541**	**638**	**641**	**633**	**634**	**659**	**629**	**651**	**636**	**636**
Wood	**2,552**	**2,874**	**2,738**	**2,534**	**2,591**	**2,628**	**2,305**	**2,211**	**2,247**	**2,317**
Total	13,314	14,543	14,630	14,514	14,655	15,057	14,540	14,232	14,456	14,673

[a]Fuel oil use in the residential sector consists of distillate fuel only. In the other sectors it includes both distillate and residual fuel oil.

*Less than 500 metric tons nitrous oxide.

P = preliminary data. See notes and sources at end of table.

Table 29. U.S. Nitrous Oxide Emissions from Stationary Combustion, 1990 and 1996-2004 (Continued)

Source	1990	1996	1997	1998	1999	2000	2001	2002	2003	P2004
	Thousand Metric Tons Nitrous Oxide									
Residential										
Coal	*	*	*	*	*	*	*	*	*	*
Fuel Oil[a]	1	1	1	1	1	1	1	1	1	1
Natural Gas	*	1	*	*	*	*	*	*	1	*
Wood	2	2	2	2	2	2	1	1	1	1
Subtotal	**4**	**4**	**3**	**3**	**3**	**3**	**3**	**3**	**3**	**3**
Commercial										
Coal	*	*	*	*	*	*	*	*	*	*
Fuel Oil[a]	1	*	*	*	*	1	1	*	*	*
Natural Gas	*	*	*	*	*	*	*	*	*	*
Wood	*	*	*	*	*	*	*	*	*	*
Subtotal	**1**	**1**	**1**	**1**	**1**	**1**	**1**	**1**	**1**	**1**
Industrial										
Coal	4	3	3	3	3	3	3	3	3	3
Fuel Oil[a]	5	6	6	6	6	6	6	6	6	6
Natural Gas	1	1	1	1	1	1	1	1	1	1
Wood	6	7	7	6	6	6	6	5	5	6
Subtotal	**16**	**17**	**17**	**16**	**17**	**16**	**16**	**15**	**15**	**15**
Electric Power										
Coal	23	26	27	27	27	28	28	28	28	29
Fuel Oil[a]	1	*	1	1	1	1	1	1	1	1
Natural Gas	*	*	*	*	*	1	1	1	1	1
Wood	1	1	1	1	1	1	*	1	1	1
Total	**24**	**27**	**28**	**29**	**29**	**30**	**30**	**30**	**30**	**30**
Total All Sectors										
Coal	**27**	**29**	**30**	**30**	**30**	**32**	**31**	**31**	**31**	**32**
Fuel Oil[a]	**8**	**8**	**8**	**8**	**8**	**8**	**8**	**8**	**8**	**8**
Natural Gas	**2**	**2**	**2**	**2**	**2**	**2**	**2**	**2**	**2**	**2**
Wood	**9**	**10**	**9**	**9**	**9**	**9**	**8**	**7**	**8**	**8**
Total	**45**	**49**	**49**	**49**	**50**	**51**	**49**	**48**	**49**	**50**

[a]Fuel oil use in the residential sector consists of distillate fuel only. In the other sectors it includes both distillate and residual fuel oil.

P = preliminary data.

Notes: Data in this table are revised from the data contained in the previous EIA report, *Emissions of Greenhouse Gases in the United States 2003,* DOE/EIA0573(2003) (Washington, DC, December 2004). Totals may not equal sum of components due to independent rounding.

Sources: Emissions coefficients from Intergovernmental Panel on Climate Change, *Greenhouse Gas Inventory Reference Manual: Revised 1996 IPCC Guidelines forNational Greenhouse Gas Inventories,* Vol. 3 (Paris, France, 1997), p. 1.50, web site www.ipcc.ch/pub/guide.htm. Energy consumption data from Energy Information Administration, *State EnergyData Report 1998,* DOE/EIA021 4(98) (Washington, DC, September2003); and *Monthly Energy Review,* DOE/EIA0035(2003/08) (Washington, DC, August 2003).

Table 30. U.S. Nitrous Oxide Emissions from Industrial Sources, 1990 and 1996-2004

Source	1990	1996	1997	1998	1999	2000	2001	2002	2003	P2004
Million Metric Tons Carbon Dioxide Equivalent										
Adipic Acid										
Controlled Sources	1.0	1.1	1.4	1.5	1.5	1.6	1.4	1.6	1.6	1.6
Uncontrolled Sources	15.9	19.6	6.4	2.0	2.1	2.1	2.1	2.3	1.4	1.4
Subtotal	**16.8**	**20.7**	**7.8**	**3.5**	**3.6**	**3.7**	**3.5**	**3.9**	**3.0**	**3.0**
Nitric Acid	**11.7**	**13.6**	**13.9**	**13.7**	**13.2**	**12.9**	**10.4**	**11.3**	**11.0**	**10.9**
Total Known Industrial Sources	**28.6**	**34.3**	**21.8**	**17.2**	**16.8**	**16.6**	**14.0**	**15.2**	**14.0**	**14.0**
Thousand Metric Tons Nitrous Oxide										
Adipic Acid										
Controlled Sources	3	4	5	5	5	5	5	5	5	6
Uncontrolled Sources	54	66	22	7	7	7	7	8	5	5
Subtotal	**57**	**70**	**27**	**12**	**12**	**13**	**12**	**13**	**10**	**10**
Nitric Acid	**40**	**46**	**47**	**46**	**45**	**43**	**35**	**38**	**37**	**37**
Total Known Industrial Sources	**96**	**116**	**74**	**58**	**57**	**56**	**47**	**51**	**47**	**47**

P = preliminary data.

Note: Data in this table are revised from the data contained in the previous EIA report, *Emissions of Greenhouse Gases in the United States 2003,* DOE/EIA0573(2003) (Washington, DC, December 2004). Totals may not equal sum of components due to independent rounding.

Sources: Data sources and methods documented in Energy Information Administration, *Documentation forEmissions of Greenhouse Gases in the United States 2003,* DOE/EIA0638(2003) (Washington, DC, January 2005), and *Documentation for Emissions of Greenhouse Gases in the United States 2004,* DOE/EIA0638(2004) (to be published).

OTHER GASES: HYDROFLUOROCARBONS, PERFLUOROCARBONS, AND SULFUR HEXAFLUORIDE

Overview

Total U.S. Emissions of Hydrofluorocarbons, Perfluorocarbons, and Sulfur Hexafluoride, 1990-2004

Estimated 2004 Emissions (Million Metric Tons Carbon Dioxide Equivalent)	155.9
Change Compared to 2003 (Million Metric Tons Carbon Dioxide Equivalent)	13.6
Change from 2003 *(Percent)*	*9.6%*
Change Compared to 1990 (Million Metric Tons Carbon Dioxide Equivalent)	67.9
Change from 1990 *(Percent)*	*77.1%*

U.S. emissions of hydrofluorocarbons (HFCs), perfluorocarbons (PFCs), and sulfur hexafluoride (SF_6) in 2004, according to the U.S. Environmental Protection Agency (EPA), totaled 155.9 million metric tons carbon dioxide equivalent ($MMTCO_2e$). Collectively, these "other gases" represented 2.2 percent of total U.S. greenhouse gas emissions. Their 2004 emissions were 9.6 percent (13.6 $MMTCO_2e$) above the 2003 level of 142.4 $MMTCO_2e$, an increase attributed primarily to a 12percent (13.5 $MMTCO_2e$) increase in emissions of HFCs.

Table 31 at the end of this chapter shows U.S. emissions of HFCs, PFCs, and SF6 from 1990 to 2004 in carbon dioxide equivalent units, and Table 32 shows their emissions in metric tons of native gas. U.S. emissions of HFCs, PFCs, and SF6 were 77 percent (67.9 $MMTCO_2e$) higher in 2004 than in 1990 (88.1 $MMTCO_2e$). Revised EPA data for 19902003 and new estimates for 2004 show that annual emissions of HFCs have increased significantly since 1990; emissions of PFCs have declined by twothirds; and SF6 emissions have declined by about onehalf since 1990 (Figure 4).

The increase in HFC emissions since the early 1990s reflects the use of HFCs as replacements for CFCs (chlorofluorocarbons), HCFCs (hydrochlorofluorocarbons), halons, and other ozonedepleting substances (ODS) that are being phased out under the Montreal Protocol because they damage the Earth's stratospheric ozone layer. ODS are used in a variety of applications, including refrigeration and air conditioning, solvents, foam production, fire extinguishers, aerosols, and sterilization.

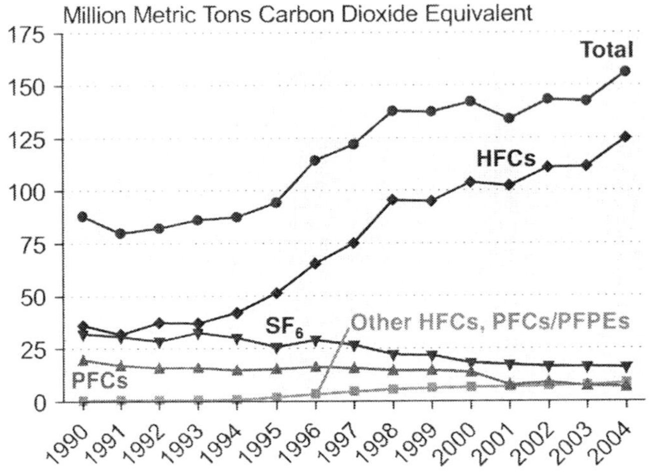

Source: Estimates presented in this chapter.

Figure 4. U.S. Emissions of Hydrofluorocarbons, Perfluorocarbons, and Sulfur Hexafluoride, 1990-2004

PFC emissions have trended downward since 1990, largely as a result of reduced PFC emissions from the aluminum industry. SF6 emissions have also declined since 1990, mainly due to reduced emissions from electricity transmission and distribution facilities in the electric power industry.

The emissions estimates in Table 31 are based on data provided by the EPA's Office of Air and Radiation.[88] The EPA data, provided in units of native gas (thousand metric tons), were converted to carbon dioxide equivalent units by the Energy Information Administration (EIA), using values for the global warming potential (GWP) for each gas from the 2001 Third Assessment Report of the Intergovernmental Panel on Climate Change (IPCC). The estimates in Table 32 are taken directly from data supplied by the EPA's Office of Air and Radiation. The 2004 preliminary estimates developed by the EPA and provided to EIA include some revisions to historical emissions estimates, based on recent runs of the EPA's Vintaging

Model (see boxes on pages 69 and 70). Those revisions are reflected in the emissions estimates presented in this chapter.

Hydrofluorocarbons (HFCs)

U.S. Emissions of Hydrofluorocarbons, 1990-2004

Estimated 2004 Emissions (Million Metric Tons Carbon Dioxide Equivalent)	124.8
Change Compared to 2003 (Million Metric Tons Carbon Dioxide Equivalent)	13.5
Change from 2003 *(Percent)*	*12.1%*
Change Compared to 1990 (Million Metric Tons Carbon Dioxide Equivalent)	88.7
Change from 1990 *(Percent)*	*246.1%*

The EPA estimates U.S. emissions of HFCs in 2004 at 124.8 $MMTCO_2e$, equivalent to 1.8 percent of total U.S. greenhouse gas emissions.[89] HFC emissions in 2004 were 12 percent (13.5 $MMTCO_2e$) above the 2003 level of 111.3 $MMTCO_2e$ (Table 31). The overall increase included increases in emissions of HFC134a (5.2 $MMTCO_2e$), HFC23 (3.4 $MMTCO_2e$), HFC143a (3.0 $MMTCO_2e$), and HFC125 (1.9 $MMTCO_2e$).

U.S. HFC emissions in 2004 were 246 percent (88.7 $MMTCO_2e$) above the 1990 level of 36.1 $MMTCO_2e$. Since 1990, HFC emissions have accounted for a growing share of total emissions of HFCs, PFCs, and SF6 combined (80 percent in 2004, compared with 41 percent in 1990). By far the largest portion of HFC emissions, 87 percent, can be attributed to their use as replacements for ODS. Emissions of HFCs used as substitutes for ODS (HFC125, HFC134a, HFC143a, and HFC236fa) have grown from trace amounts in 1990 to 108.5 $MMTCO_2e$ in 2004.

HFCs are compounds containing carbon, hydrogen, and fluorine. Although they do not destroy stratospheric ozone, they are powerful greenhouse gases. HFCs are used in many applications, such as solvents, domestic and commercial refrigerants, firefighting agents, propellants for pharmaceutical and industrial aerosols, foamblowing agents, and in blends for air conditioning refrigerants.

The market for HFCs is expanding. As CFCs and ODS are being phased out under the Montreal Protocol and the Clean Air Act, HFCs have been introduced into the market to fill the void in many key applications. For example, HFCs are used in fire protection applications to replace Halon 1301 and Halon 1211, which are no longer being produced in the United States.[90] HCFCs, now interim replacements for CFCs, will also be phased out. For example, HCFC141b and HCFC142b, which are used as blowing agents in insulation foams, will be replaced by HFCs for some uses.[91]

Trifluoromethane (HFC23)

The EPA estimates 2004 emissions of HFC23 at 16.3 $MMTCO_2e$.[92] HFC23 emissions, representing 13 percent of total HFC emissions in 2004, were 26 percent (3.4 $MMTCO_2e$) above their 2003 level of 13.0 $MMTCO_2e$— the first upturn in HFC23 emissions since 1998—but still 55 percent (19.7 $MMTCO_2e$) below their 1990 level of 36.1 $MMTCO_2e$. Since 1990, annual HFC23 emissions have fluctuated, peaking in 1998 at 41.7 $MMTCO_2e$ and then falling steadily until the upturn in 2004.

As in 1998, the 2004 increase in HFC23 emissions can be traced to an increase in the production of chlorodifluoromethane (HCFC22). Nearly all HFC23 emissions (98 percent) are created as a byproduct in the production of HCFC22 and generally are vented to the atmosphere. In some cases the HFC23 is captured for use in a limited number of applications. While production of HCFC22 peaked in 2000, emissions of HFC23 from this source declined from 1998 until 2004, because the HFC23 emission rate (i.e., the amount of HFC23 emitted per kilogram of HCFC22 manufactured) decreased significantly.[93] Production of HCFC22 also fell significantly between 2000 and 2004, resulting in further reductions in emissions.[94]

HCFC22 is used as a component of blowing agents for polyurethane foams and extruded polystyrene foams, and in the refrigerant market for stationary refrigeration and air conditioning (including chillers, room and household [central] air conditioners, and dehumidifiers). The EPA administers a voluntary program (the HFC23 Emission Reduction Program) with HCFC22 producers to reduce HFC23 emissions, which has helped to moderate HFC23 emissions during periods of rising demand for HCFC22. In the long term, domestic production of HCFC22 for nonfeedstock uses will be phased out by 2020 under the U.S. Clean Air Act, pursuant to U.S. agreements under the Copenhagen Amendments to the Montreal Protocol. However, HCFC22 production for use as a feedstock in the production of other chemicals (fluorinated polymers) will be allowed to continue indefinitely.[95]

Revisions in EPA Emissions Estimation Methodology

The primary source for the emissions estimates presented in this chapter is data obtained from the U.S. Environmental Protection Agency (EPA), Office of Air and Radiation. The Office of Air and Radiation also prepares an annual inventory of greenhouse gas emissions, which is published pursuant to U.S. commitments under the United Nations Framework Convention on Climate Change (UNFCCC). The UNFCCC encourages parties to revise methods regularly and to recaculate emissions affected by the revisions. The data supporting the EPA inventory, including the emissions estimates for 2004, incorporate a number of revisions to the data and estimation methodologies used for hydrofluorocarbons (HFCs), perfluorocarbons (PFCs), and sulfur hexafluoride (SF_6) in its most recent emissions inventory.[a] Those changes are reflected in the EPA's historical emissions estimates, as described below:

- Electricity Transmission and Distribution. The changes in calculations of emissions from electricity transmission and distribution are the result of incorporating more uptodate transmission mileage data and the inclusion of additional historical partner data in the EPA's SF_6 Emissions Reduction Partnership for Electric Power Systems for 2000, 2001, and 2002. Previously, the 2001 Utility Data Institute (UDI) Directory of Electric Power Producers and Distributors was used by the EPA to estimate SF_6 emissions for 2001 and 2002. Those numbers have been revised to account for increases in transmission mileage during 2001 and 2002, primarily as a result of growth in the U.S. transmission system. Accordingly, estimates of nonpartner and nonreporting partner emissions have been recalculated in the nonreporting partner regression equations. Because transmission miles are highly correlated with SF_6 emissions, the EPA has used these regression equations to calculate emissions from nonpartners and nonreporting partners in the SF_6 emissions reduction partnership. In addition to transmission mileage revisions, the electric power system emission estimates have also been recalculated, based on additional historical partner data. Specifically, the regression equations for each respective year of the historical partner submissions have been updated, resulting in new extrapolations to nonreporting partners as well. These revisions resulted in an average annual decrease in estimated SF_6 emissions from electric power systems of 0.2 percent, or less than 0.1 million metric tons carbon dioxide equivalent ($MMTCO_2e$) for the 20002002 period.

- Magnesium Production and Processing. The emissions estimates in this report have been revised to reflect new historical data supplied by the U.S. Geological Survey and participants in the EPA's SF_6 Emission Reduction Partnership for the Magnesium Industry. This change resulted in an average annual increase in estimated SF_6 emissions from magnesium production and processing of less than 0.1 $MMTCO_2e$ (4.1 percent) for the 20002002 period
- Substitution of OzoneDepleting Substances. The EPA has updated assumptions for its Vintaging Model pertaining to market trends in chemicals and chemical substitutes. These changes resulted in an average annual net increase in estimated HFC and PFC emissions of less than 0.1 $MMTCO_2e$ (4.1 percent) for the 19902002 period.
- Aluminum Production. As the result of an EPAfunded study, facilityspecific slope coefficients for three U.S. aluminum smelters have been reestimated. The new coefficients have been used by the EPA in place of the IPCC defaults for revising the appropriate smelterspecific emission factors. The EPA provided the revised data to EIA, along with additional recently reported data concerning smelter operating parameters by participants in the EPA's Voluntary Aluminum Industrial Partnership Program. These changes resulted in an average annual increase of less than 0.1 $MMTCO_2e$ (0.2 percent) for the 19902002 period.
- HCFC22 Production. Based on conversations with the Alliance for Responsible Atmospheric Policy, the EPA has adjusted the historical time series for HFC emissions from HCFC22 production. These changes resulted in an average annual decrease in HFC emissions from HCFC22 production of less than 0.1$MMTCO_2e$ (0.01 percent) for the 19902002 period.

The U.S. Environmental Protection Agency (EPA) uses a detailed Vintaging Model for equipment and products containing ozonedepleting substances (ODS) and ODS substitutes to estimate actual versus potential emissions of various ODS substitutes, including hydrofluorocarbons (HFCs) and perfluorocarbons (PFCs). The model estimates the quantities of equipment and products sold each year that contain ODS and ODS substitutes, and the amounts of chemicals required for their manufacture and/or maintenance over time.

Emissions from more than 40 different end uses are estimated by applying annual leak rates and release profiles, which account for the lag in emissions from equipment as it leaks over time.

For most products (refrigerators, air conditioners, fire extinguishers, etc.), emissions calculations are split into two categories: emissions during equipment lifetime, which arise from annual leakage and service losses plus emissions from manufacture; and disposal emissions, which occur when the equipment is discarded. By aggregating the data over different end uses, the model produces estimates of annual use and emissions of each compound [a].

The EPA periodically attempts to improve the model and reduce the uncertainty of emissions estimates by using more accurate data from emitting industries. The level of detail incorporated in the EPA Vintaging Model is higher than that of the default methodology used by the Intergovernmental Panel on Climate Change, although there still is some uncertainty about some of the model inputs, such as equipment characteristics and sales figures.

Tetrafluoroethane (HFC134a)

The EPA estimates 2004 U.S emissions of HFC134a at 65.8 $MMTCO_2e$.[96] HFC134a accounts for the largest share of total HFC emissions (53 percent). The 2004 emissions level is 8.7 percent (5.2 $MMTCO_2e$) higher than the 2003 level of 60.5 $MMTCO_2e$. The increase can be attributed primarily to the continued use of HFC134a as a substitute for CFCs in motor vehicle air conditioning systems.

Annual HFC134a emissions have grown dramatically from their estimated 1992 level of 0.7 $MMTCO_2e$. Since 1994, HFC134a has been the transportation industry standard for replacing CFCs in air conditioners for passenger cars, trucks, trains, and buses, because it is nonflammable, has low toxicity, and is not an ODS.

HFC134a is also used in refrigerant blends (e.g., R404A) in most new commercial refrigeration equipment built in the United States and in commercial chillers, but leakage from these sources is much less than from automotive air conditioners. Leakage occurs primarily during the servicing of the units rather than during normal operation. Shortterm uses of HFC134a, on the other hand, are becoming an important source of emissions. Such uses include aerosols and opencell foam blowing, which are denoted as shortterm uses because most of the HFC134a used will be emitted to the atmosphere within a short period of time. According to the Alternative Fluorocarbons Environmental Acceptability Study (AFEAS), worldwide sales of HFC134a jumped more than fourfold between 1992 and 1993, doubled again in 1994, and continued growing steadily to 166,899 metric tons of gas in 2003.[97]

Pentafluoroethane (HFC125)

The EPA estimates U.S. emissions of HFC125 in 2004 at 19.7 $MMTCO_2e$, equivalent to 16 percent of total HFC emissions.[98] The 2004 emissions level is 10 percent (1.9

$MMTCO_2e$) higher than the 2003 level of 17.8 $MMTCO_2e$. Emissions of HFC125 have increased steadily from 0.7 $MMTCO_2e$ in 1992 because of its use as a refrigerant blending agent.

HFC125 is used in the blend R410A, which is designed to replace HCFC22 as the refrigerant of choice for stationary commercial refrigeration and air conditioning applications, as well as in the blends R404A and R507A. Some manufacturers have already introduced air conditioners that use R410A, but as yet the product has captured only a small percentage of the market. As the phaseout of HCFC22 begins to gain momentum, producers expect a rapid increase in demand for R410A.[99] HFC125 can also be used as a firefighting agent.

Trifluoroethane (HFC143a)

The EPA estimates U.S. emissions of HFC143a in 2004 at 19.5 MMTCO_2e, equivalent to 16 percent of total HFC emissions.[100] The 2004 emissions level is 18 percent (3.0 MMTCO_2e) higher than the 2003 level of 16.5 MMTCO_2e. HFC143a emissions have increased rapidly from 0.1 MMTCO_2e in 1993, as demand for HFC143a as a refrigerant blending agent has increased.

HFC143a is a halocarbon used in blends for commercial refrigeration and air conditioning, such as R404A and R507A. HFC143a, like other HFCs, is used as a substitute because it contains neither chlorine nor bromine and does not emit ozonedepleting halogen radicals into the stratosphere. Like other halocarbons, HFC143a does make a positive contribution to atmospheric warming; however, the GWPs of R404A and R507A are lower than those of the gases it replaces, such as CFC12 with a GWP of 10,600.

Hexafluoropropane (HFC236fa)

The EPA estimates U.S. emissions of HFC236fa in 2004 at 3.5 MMTCO_2e, equivalent to 2.8 percent of total HFC emissions.[101] The 2004 emissions level is essentially unchanged from the 2003 level of 3.5 MMTCO_2e. Emissions of HFC236fa have increased from 0.1 MMTCO_2e in 1997 because of its use as a refrigerant, in particular by the U.S. Navy for shipboard applications.[102] In another application, HFC236fa is used as a firefighting agent.

Perfluorocarbons (PFCs)

U.S. Emissions of Perfluorocarbons, 1990-2004

Estimated 2004 Emissions (Million Metric Tons Carbon Dioxide Equivalent)	6.7
Change Compared to 2003 (Million Metric Tons Carbon Dioxide Equivalent)	-0.5
Change from 2003 *(Percent)*	*-7.5%*
Change Compared to 1990 (Million Metric Tons Carbon Dioxide Equivalent)	-13.0
Change from 1990 *(Percent)*	*-66.1%*

The EPA estimates 2004 emissions of PFCs at 6.7 MMTCO_2e, accounting for 4.3 percent of all emissions of HFCs, PFCs, and SF6 combined. The estimate for 2004 is 7.5 percent (0.5 MMTCO_2e) lower than the estimate for 2003 (7.2 MMTCO_2e) and 66 percent (13.0 MMTCO_2e) lower than the 1990 emissions level of 19.6 MMTCO_2e.[103] The downward trend in emissions is largely the result of decreases in domestic aluminum production, which creates PFCs as byproducts, as well as process efficiency improvements in the aluminum industry. Moderating the decrease in PFC emissions from aluminum manufacture have been periodic increases in PFC emissions from semiconductor manufacture. The two PFCs emitted by these industries are perfluoromethane (CF_4) and perfluoroethane (C_2F_6).

Perfluoromethane (CF_4)

The EPA estimates U.S. emissions of CF4 in 2004 at 3.2 $MMTCO_2e$, equivalent to 48 percent of total PFC emissions.[104] The 2004 emissions level is 16 percent (0.6 $MMTCO_2e$) lower than the 2003 level of 3.8 $MMTCO_2e$. CF4 emissions have dropped by a total of 79 percent (11.7 $MMTCO_2e$) from their 1990 level of 14.8 $MMTCO_2e$.

The two principal sources of CF_4, as well as C2F6, are as a byproduct of aluminum smelting created during periods of process inefficiency and disruption, and in the manufacture of semiconductors. The EPA estimates U.S. CF4 emissions in 2004 from aluminum production at 2.1 $MMTCO_2e$ and from semiconductor manufacture at 1.1 $MMTCO_2e$.[105] With reductions in primary aluminum production and improvements that reduce anode effects leading to process inefficiency, CF4 emissions from aluminum smelting have been reduced by 85 percent (12.1 $MMTCO_2e$) from their 1990 level of 14.2 $MMTCO_2e$.

Aluminum smelting companies that participate in the EPA's Voluntary Aluminum Industry Partnership (VAIP) have achieved efficiency improvements through voluntary actions. Reductions in primary aluminum production have also played a role in reducing PFC emissions. According to data from the U.S. Geological Survey (USGS), domestic primary aluminum production decreased significantly between the years 2000 and 2003 and, according to preliminary USGS estimates, fell again in 2004. The decline in production resulted from cutbacks in smelter production in response to higher costs of energy and alumina. Most of the production cutbacks took place in the Pacific Northwest.[106]

Another source of CF4 emissions is semiconductor manufacturing. Emissions from this source peaked in 2000 at just over 1.6 $MMTCO_2e$ and have since declined by 34 percent (0.6 $MMTCO_2e$) to their 2004 level of 1.1 $MMTCO_2e$.[107] This estimate reflects the rapid growth of the semiconductor industry in the 1990s, which has resulted in a 61percent increase in emissions (0.4 $MMTCO_2e$) from their 1990 level of 0.7 $MMTCO_2e$. CF_4, like C2F6, is used as a plasma etchant and cleaning agent in semiconductor manufacturing; some of the gas used in those processes does not react with the materials and, unless abated, is emitted to the atmosphere. A variety of other perfluorinated compounds are used in the semiconductor industry, including perfluoropropane (C_3F_8, with a GWP of 8,600), perfluorobutane (C4F10, GWP 8,600), perfluorohexane (C6F14, GWP 9,000), and nitrogen trifluoride (NF_3).[108]

Perfluoroethane (C2F6)

The EPA estimates U.S. emissions of C2F6 in 2004 at 3.5 $MMTCO_2e$, equivalent to 52 percent of total PFC emissions.[109] The 2004 emissions level is 2.4 percent (0.1 $MMTCO_2e$) above the 2003 level of 3.4 $MMTCO_2e$. Emissions of C2F6 have dropped by 28 percent (1.3 $MMTCO_2e$) from their 1990 level of 4.8 $MMTCO_2e$.

C2F6 emissions originate from the aluminum and semiconductor industries. Thus, C2F6, like CF_4, has had two countervailing trends bearing on its emissions levels. On one side, decreasing aluminum production and increased efficiency in the aluminum industry have tended to lower emissions. C2F6 emissions in the aluminum industry fell from 2.9 $MMTCO_2e$ in 1990 to 0.6 $MMTCO_2e$ in 2004, or by 81 percent (2.3 $MMTCO_2e$). On the other side, increased semiconductor production has tended to increase emissions. C2F6 emissions in the semiconductor industry, estimated by the EPA at 2.9 $MMTCO_2e$ in 2004, have increased by 53 percent (1.0 $MMTCO_2e$) from their 1990 level of 1.9 $MMTCO_2e$. The net effect has been a 28percent overall reduction in emissions of C_2F_6 since 1990.

Other HFCs and PFCs/PFPEs

There is a group of other HFCs and PFCs/PFPEs for which the EPA withholds individual emissions data, because the data are considered confidential and could compromise business practices. This group includes HFC152a, HFC227ea, HFC245fa, and HFC4310mee.[110] The EPA estimates total emissions of this group of "other HFCs" at 8.6 $MMTCO_2e$ in 2004, representing 5.5 percent of all emissions of HFCs, PFCs, and SF6 reported.[111] Emissions of these HFCs are small but growing rapidly, as they continue to find applications as substitutes for CFCs and HCFCs. Emissions of "other HFCs" increased by 13 percent (1.0 $MMTCO_2e$) in 2004 compared with 2003 (7.6 $MMTCO_2e$).

Other HFCs and HFC blends are also likely to gain market share as a result of the phaseout of CFCs and HCFCs, because no single product is suited for all applications. For example, each potential replacement product has an optimal operating temperature range; hence, the refrigerant best suited for use in ice cream freezers will differ from the best choice for milk coolers.[112]

In addition to replacing HCFC22 in stationary air conditioning and refrigeration applications, other HFCs are expected to gain new markets as foamblowing agents. CFCs have already been phased out of this market, having been replaced by HCFCs (primarily HCFC141b). Among the potential replacements for HCFCs, HFC245fa (pentafluoropropane) appears to be the strongest contender.[113]

Sulfur Hexafluoride (SF_6)

U.S. Emissions of Sulfur Hexafluoride, 1990-2004

Estimated 2004 Emissions (Million Metric Tons Carbon Dioxide Equivalent)	16.0
Change Compared to 2003 (Million Metric Tons Carbon Dioxide Equivalent)	-0.4
Change from 2003 *(Percent)*	*-2.2%*
Change Compared to 1990 (Million Metric Tons Carbon Dioxide Equivalent)	-16.1
Change from 1990 *(Percent)*	*-50.2%*

The EPA estimates 2004 emissions of SF6 at 16.0 $MMTCO_2e$, accounting for 10 percent of all HFC, PFC, and SF6 emissions combined in 2004.[114] Emissions in 2004 were 2.2 percent (0.4 $MMTCO_2e$) lower than in 2003 (16.3 $MMTCO_2e$) and 50 percent (16.1

$MMTCO_2e$) lower than the estimate for 1990 (32.1 $MMTCO_2e$). The downward trend in SF6 emissions since 1990 is the result of industry efforts to reduce emissions from electrical power systems, as well as the rising cost of SF_6. In contrast, emissions of SF6 from uses in the semiconductor manufacturing industry have increased overall by 76 percent since 1990.

SF6 is used primarily in electrical applications, in which it is an excellent dielectric gas for highvoltage applications, because it is chemically inert, gaseous at low temperatures, nonflammable, nontoxic, and noncorrosive.[115] In electricity transmission and distribution systems, SF6 acts as an insulator and arc interrupter for circuit breakers, switch gear, and other electrical equipment; however, it can escape through seals, especially in older equipment. Emissions also occur during equipment installation, servicing, and disposal.[116]

Other applications that produce SF6 emissions include magnesium metal casting processes that employ SF6 to replace toxic and corrosive materials, such as salt fluxes and sulfur dioxide (SO_2). Another use of SF6 is as a cover gas during magnesium production and processing to prevent excessive oxidation of molten magnesium in the presence of air, although emissions from this source have declined due to a decrease in the quantity of magnesium die casting, process optimizations by industry participants in EPA's SF6 Emission Reduction Partnership for the Magnesium Industry, and the closure of primary magnesium processing facilities in the United States in 2001.[117] Pretreating aluminum melt with SF6 (or an inert gas mixture) prevents porosity and therefore weakening of the metal. It also removes oxides and solid impurities. In addition, mixtures of SF6 and O2 are used as feed gases for plasma etching of semiconductor devices.[118] Because of its extremely low atmospheric concentration, SF6 is also useful as an atmospheric tracer gas for a variety of experimental purposes. Other minor applications include leak detection, loudspeakers, lasers, and as a cover gas or fluxing and degassing agent for specialized casting operations in the aluminum industry.[119]

Table 31. U.S. Emissions of Hydrofluorocarbons, Perfluorocarbons, and Sulfur Hexafluoride, 1990 and 1996-2004

Gas	1990	1996	1997	1998	1999	2000	2001	2002	2003	P2004
Hydrofluorocarbons										
HFC-23	36.1	32.3	31.2	41.7	31.7	30.9	20.6	20.6	13.0	16.3
HFC-125	0.0	6.3	8.6	10.7	12.1	13.6	14.9	16.2	17.8	19.7
HFC-134a	0.0	24.5	31.4	36.7	42.2	48.0	52.7	56.9	60.5	65.8
HFC-143a	0.0	2.3	4.0	5.9	7.5	9.3	11.4	13.8	16.5	19.5
HFC-236fa	0.0	0.0	0.1	0.6	1.3	2.0	2.6	3.2	3.5	3.5
Total HFCs	**36.1**	**65.5**	**75.1**	**95.5**	**94.9**	**103.8**	**102.2**	**110.7**	**111.3**	**124.8**
Perfluorocarbons										
CF_4	14.8	11.0	10.1	8.7	8.6	8.7	4.2	5.0	3.8	3.2
C_2F_6	4.8	5.4	5.6	5.9	6.0	5.0	3.4	3.8	3.4	3.5
Total PFCs	**19.6**	**16.4**	**15.7**	**14.6**	**14.6**	**13.7**	**7.6**	**8.8**	**7.2**	**6.7**
Other HFCs, PFCs/PFPEs	**0.3**	**3.4**	**4.5**	**5.5**	**6.2**	**6.5**	**6.7**	**7.1**	**7.6**	**8.6**
Sulfur Hexafluoride	**32.1**	**29.1**	**26.7**	**22.1**	**21.7**	**18.1**	**17.4**	**16.4**	**16.3**	**16.0**
Total Emissions	**88.1**	**114.3**	**122.0**	**137.7**	**137.4**	**142.1**	**133.9**	**143.1**	**142.4**	**155.9**

P = preliminary data.

Notes: Other HFCs, PFCs/PFPEs include HFC152a, HFC227ea, HFC245fa, HFC4310mee, and a variety of PFCs and perfluoropolyethers (PFPEs). They are grouped together to protect confidential data. Totals may not equal sum of components due to independent rounding.

Source: U.S. Environmental Protection Agency, Office of Air and Radiation, web site www.epa.gov/globalwarming/ (preliminary estimates, September 2005).

Table 32. U.S. Emissions of Hydrofluorocarbons, Perfluorocarbons, and Sulfur Hexafluoride, 1990 and 1996-2004

Gas	1990	1996	1997	1998	1999	2000	2001	2002	2003	P2004
Hydrofluorocarbons										
HFC-23	3.0	2.7	2.6	3.5	2.6	2.6	1.7	1.7	1.1	1.4
HFC-125	0.0	1.9	2.5	3.1	3.6	4.0	4.4	4.8	5.2	5.8
HFC-134a	0.0	18.9	24.1	28.2	32.5	36.9	40.5	43.8	46.6	50.6
HFC-143a	0.0	0.5	0.9	1.4	1.7	2.2	2.6	3.2	3.8	4.5
HFC-236fa	0.0	0.0	*	0.1	0.1	0.2	0.3	0.3	0.4	0.4
Perfluorocarbons										
CF_4	2.6	1.9	1.8	1.5	1.5	1.5	0.7	0.9	0.7	0.6
C_2F_6	0.4	0.4	0.5	0.5	0.5	0.4	0.3	0.3	0.3	0.3
Other HFCs, PFCs/PFPEs	M	M	M	M	M	M	M	M	M	M
Sulfur Hexafluoride	**1.4**	**1.3**	**1.2**	**1.0**	**1.0**	**0.8**	**0.8**	**0.7**	**0.7**	**0.7**

*Less than 50 metric tons of gas.

P = preliminary data. M = mixture of gases.

Notes: Other HFCs, PFCs/PFPEs include HFC152a, HFC227ea, HFC245fa, HFC4310mee, and a variety of PFCs and perfluoropolyethers (PFPEs). They are grouped together to protect confidential data. Totals may not equal sum of components due to independent rounding.

Source: U.S. Environmental Protection Agency, Office of Air and Radiation, web site www.epa.gov/globalwarming/ (preliminary estimates, September 2005).

LAND-USE ISSUES

Overview

Land-use and forestry issues are important to national and global inventories of greenhouse gases in three ways:

Vegetation can "sequester" or remove carbon dioxide from the atmosphere and store it for potentially long periods in above and belowground biomass, as well as in soils. Soils, trees, crops, and other vegetation may make significant contributions to reducing net greenhouse gas emissions by serving as carbon "sinks."

Harvested wood put into wood products, or eventually into landfills, can potentially sequester carbon dioxide from the atmosphere for decades before the carbon stored in the wood products decays and is released to the atmosphere.

Humaninduced land-use changes and forest management practices can alter the quantities of atmospheric and terrestrial carbon stocks, as well as the natural carbon flux among biomass, soils, and the atmosphere.[120]

Land-use issues are of particular interest to U.S. policymakers, because U.S. forests and soils annually sequester large amounts of carbon dioxide. Much of the forest land in the United States was originally cleared for agriculture, lumber, or fuel in the hundred years before 1920. Since then, however, much of the agricultural and pasture land has reverted to forest land, increasing its ability to sequester atmospheric carbon dioxide.

The amount of carbon being sequestered annually is uncertain, in part because of an absence of data and difficulties in measuring carbon sequestration. Moreover, in addition to technical uncertainties, there are also policy and accounting questions about the aspects of the carbon cycle that should be included in national inventories as anthropogenic emissions and removals.

The 1996 revised guidelines for national emissions inventories, published in 1997 by the Intergovernmental Panel on Climate Change (IPCC), include methods for calculating carbon sequestration and net carbon dioxide flux to the atmosphere resulting from land-use changes and land-use activities, such as forestry.[121] The U.S. Environmental Protection Agency (EPA) estimates annual U.S. carbon sequestration in 2003, based on data generated by the U.S. Department of Agriculture (USDA), at 828.0 million metric tons carbon dioxide equivalent ($MMTCO_2e$), a decline of approximately 21 percent from the 1,042.1 $MMTCO_2e$ sequestered in 1990[122] (Table 33). Land use, land-use change,

Table 33. Net Carbon Dioxide Sequestration from U.S. Land-use Change and Forestry, 1990 and 19972003

Component	1990	1997	1998	1999	2000	2001	2002	2003
Forest Land Remaining Forest Land: Changes in Forest Carbon Stocks	949.3	851.0	805.5	751.7	747.9	750.9	751.5	752.7
Cropland Remaining Cropland: Changes in Agricultural Soil Carbon Stocks	8.1	7.4	4.3	4.3	5.7	7.1	6.2	6.6
Settlements Remaining Settlements	84.7	71.6	71.2	70.0	68.9	68.9	68.8	68.7
Urban Trees	*58.7*	*58.7*	*58.7*	*58.7*	*58.7*	*58.7*	*58.7*	*58.7*
Landfilled Yard Trimmings and Food Scraps	*26.0*	*12.9*	*12.5*	*11.4*	*10.2*	*10.3*	*10.2*	*10.1*
Total Net Flux	**1,042.1**	**930.0**	**881.0**	**826.0**	**822.5**	**826.9**	**826.5**	**828.0**

Note: Totals may not equal sum of components due to independent rounding.

Source: U.S. Environmental Protection Agency, *Inventory of U.S. Greenhouse Gas Emissions and Sinks: 19902003,* EPA 430R05003 (Washington, DC, April 2005), web site http://yosemite.epa.gov/oar/globalwarming.nsf/content/ResourceCenter PublicationsGHGEmissionsUSEmissionsInventory2005.html. and forestry practices offset approximately 16.9 percent of total U.S. anthropogenic carbon dioxide emissions in 1990 and 11.9 percent in 2003.[123]

New IPCC Good Practice Guidance for Land-Use, Land-use Change and Forestry

The estimates of carbon sequestration in this chapter involve several categorical and methodological changes from previous years, which have been implemented by the EPA in response to new land-use guidelines issued by the IPCC in 2003. The IPCC *Good Practice Guidance for Land Use, Land-use Change and Forestry*[124] (LULUCF GPG) recommends reporting carbon stocks according to several land-use types and conversions—for example forest land remaining forest land, nonforest land becoming forest, and forest land becoming nonforest land. These categories of "land-use type remaining land-use type" and "land-use type becoming land-use type" are a new convention adopted in the LULUCF GPG.

Currently, there are no consistent datasets for the entire United States that would allow the results for forest lands, croplands, and settlements to be disaggregated in this fashion. Thus, the net changes in carbon sequestered are aggregated to one category for each land-use type: forests, croplands, and settlements. For example, in the case of forest lands, net changes in forest carbon stocks for "forest land remaining forest land" encompass all forestrelated land, including nonforest land converted to forest and forest land converted to nonforest land.

The EPA report, *Inventory of U.S. Greenhouse Gas Emissions and Sinks: 19902003,* upon which this chapter is based, is the first to follow the LULUCF GPG. The LULUCF

GPG had to be consistent with the 1996 IPCC Guidelines, because there was an existing agreement among Parties to the United Nations Framework Convention on Climate Change (UNFCCC) to use the latter.[125] The IPCC defines this consistency as: (1) traceability of specific sources or sinks addressed by the LULUCF GPG to categories in the IPCC Guidelines; (2) use of the same functional forms (or their equivalent) for the equations in the LULUCF GPG that are used in the IPCC Guidelines; and (3) facilitation for the correction of errors or deficiencies in the IPCC Guidelines by application of the LULUCF GPG.

In its most recent inventory, the EPA followed the LULUCF GPG, reporting fluxes according to changes within and conversions between forest lands, croplands, settlements, grasslands, and wetlands. Due to the lack of consistent datasets, the EPA limited its estimates of greenhouse gas flux to: (1) forest land remaining forest land, (2) croplands remaining croplands, and (3) settlements remaining settlements. This new categorization provides additional sources of information on nitrous oxide emissions by major land-use type. Further, the EPA cautioned that other land-use and land-use change activities cause fluxes of greenhouse gases other than carbon dioxide that are not accounted for, including methane from managed forest soils and artificially flooded lands.

Significant differences between the sequestration values in this report and those in EIA's previous reports on U.S. greenhouse gas emissions are broadly attributable to changes that have been made in order to conform with the LULUCF GPG, as well as a variety of differences in calculation methods. Definitional changes include the following:

- The forest soil pool is now termed "soil organic carbon."
- The forest floor is now termed "litter."
- Previously, the tree pool included the mass of standing dead trees; now, the mass of standing dead trees, together with down dead wood, is categorized as "dead wood."
- Previously, the remainder of the tree pool, live biomass, and understory pool was divided into aboveand belowground portions; now, the aboveground tree and understory pools are summed into an aboveground biomass pool, and the belowground portions are summed into a belowground biomass pool.

Important differences in calculation methods include the following:

- The USDA's State Soil Geographic (STATSGO) database[126] and its relationship with data from the [123]EIA does not include sequestration from land-use change and forestry as part of its annual estimate of emissions of greenhouse gases in the United States. Note that land use refers to maintaining land within a particular category of use, such as forests remaining forests, whereas land-use change refers to changing from one land-use type to another, as when forest is converted to grasslands, or wetlands drained to create more land for agriculture.
- USDA's National Forest Inventory and Analysis (FIA) survey[127] are now interpreted differently for the estimation of soil organic carbon. For this report, soil organic carbon in the conterminous United States was calculated using the STATSGO database, and data gaps were filled with representative values for similar soils. Links to regions and forest types were developed with the assistance of the USDA's FIA

Geospatial Data Services, by overlaying FIA forest inventory plots with existing soil carbon maps.

- The newer USDA Forestry Inventory and Analysis Database (FIADB) datasets were considered for nonsoil forest carbon estimates, along with the USDA Resources Planning Act (RPA) data.[128] For last year's estimates, only RPA data were used when needed.
- Final values for all carbon pools were extrapolated from the last carbon stock change values calculated from the FIA survey data, because it was not possible to model final stocks in a manner consistent with available inventory data. Previously, the estimates of final carbon stocks were based on model results.

Forest Land Remaining Forest Land

The values for forest carbon dioxide fluxes reported in this chapter are based on estimates of changes in forest carbon stocks. The components analyzed are aboveground biomass, belowground biomass, dead wood, litter, soil organic carbon, harvested wood products in use, and harvested wood products in landfills. The estimated carbon dioxide flux from each of these components was calculated using FIADB data and methodologies consistent with the LULUCF GPG and the Revised 1996 IPCC Guidelines. Nitrous oxide emissions from fertilized forest soils were calculated by using a default methodology consistent with the LULUCF GPG. Pine trees, being the dominant species planted for timber in the southeastern United States, were taken as representative of fertilized forests in the country, and the average reported fertilization rate of 150 pounds of nitrogen per acre was multiplied by the area of pine receiving fertilizer.

Croplands Remaining Croplands

Estimates of carbon fluxes from croplands include changes in agricultural soil carbon stocks on both croplands and grazing lands, because datasets necessary to separate the two were not available. Changes in agricultural soil carbon stocks result from the use and management of cropland and grazing land and emissions of carbon dioxide from the application of crushed limestone and dolomite. The estimation methods used for this report are consistent with the Revised 1996 IPCC Guidelines and the LULUCF GPG.

Settlements Remaining Settlements

Fluxes from settled lands include methane from landfilled yard trimmings and food scraps, carbon from urban trees, and nitrous oxide from fertilized soils. Changes in carbon flux were estimated by analyzing lifecycle emissions and sinks associated with waste management. Stock changes in urban trees were estimated on the basis of field measurements and data on national urban tree cover, using a methodology consistent with the LULUCF GPG to estimate carbon flux. Nitrous oxide emissions from nitrogen applied to turf grass were estimated as 10 percent of all synthetic fertilizer used in the United States.

Land-use Change and Forestry Carbon Sequestration

The EPA's estimates for carbon sequestration from land-use change and forestry in 2003 include three main sink categories: (1) changes in forest carbon stocks for forest land remaining forest land (752.7 MMTCO$_2$e or 91 percent of the total); (2) changes in agricultural soil carbon stocks for cropland remaining cropland (6.6 MMTCO$_2$e or 0.8 percent of the total); and (3) changes in settlements remaining settlements (68.7 MMTCO$_2$e or 8.3 percent, including 58.7 MMTCO$_2$e from urban trees and 10.1 MMTCO$_2$e from landfilled yard trimmings and food scraps).[129] For a discussion of worldwide trends in forest and croplands see the box on page 78.

Millennium Ecosystem Assessment: Forest and Cultivated Systems

In 2000, United Nations SecretaryGeneral Kofi Annan called for the Millennium Ecosystem Assessment (MA) in his report to the United Nations General Assembly. The global assessment was completed in 2005,[a] and work continues on subglobal assessments. The MA was carried out under the auspices of the U.N. Environment Program and was governed by a multistakeholder board of representatives of international institutions, governments, businesses, nongovernmental organizations, and indigenous peoples. More than 1,300 authors from 95 countries worked to prepare the global assessment, and hundreds are continuing to work on more than 20 subglobal assessments, to be published in 2006. In addition, 850 experts were involved in peer reviews of the global assessment.

The objectives of the assessment were to gauge the effects of ecosystem change on human wellbeing and establish a scientific foundation for enhancing the conservation and sustainable use of ecosystems. The MA responds formally to government requests for information received through four international conventions, the Convention on Biological Diversity, the United Nations Convention to Combat Desertification, the Ramsar Convention on Wetlands, and the Convention on Migratory Species. Forest and cultivated land systems were an important part of the assessment.

Forest Systems

The MA defines forest systems as areas with a canopy cover of at least 40 percent provided by woody plants taller than 5 meters, including temporarily cutover forests and plantations, but excluding orchards and agroforests that mainly produce food crops. Forest systems regulate 57 percent of total water runoff, and they ensure all or some of the water supply for about 4.6 billion people worldwide.

According to the MA, the global area of forest systems has been halved over the past three centuries; 25 countries have lost all forests; and another 29 have lost more than 90 percent of their forest cover. Forest systems regulate 57 percent of total water runoff, ensuring some or all of the water supply for about 4.6 billion people. From 1990 to 2000, temperate forests increased by almost 3 million hectares annually; however, tropical deforestation exceeded 12 million hectares annually. About 40 percent of forest area has been lost during the industrial era, and forests continue to be lost in many regions. Some of the pressure to clear forests in the future, according to the MA, will be alleviated by the expanding role of plantations in timber supply.

Global timber production has increased by 60 percent in the past four decades. Most of the increase is attributed to plantations, which produced 35 percent of the global roundwood harvest in the year 2000. This proportion, according to the MA, is expected to increase to 44 percent by 2020. The most rapid expansion of plantations is expected to occur in the midlatitudes, where yields are higher and costs are lower.

Under the MA scenarios, which run from 1970 to 2050, forest area is expected to increase in industrial regions and decrease in developing ones.b Global deforestation in three of the four scenarios is projected to be approximately equal to historic rates (approximately 0.4 percent annually between 1970 and 1995). The fourth scenario projects a deforestation rate of 0.6 percent per year. Particular ecosystems, such as tropical forests, could be subject to higher than average deforestation rates.

Cultivated Systems

In the MA, cultivated systems include predominantly cropped areas, agroforestry, and aquaculture. In the past two decades, one of the areas with the greatest expansion of cropland was the U.S. Great Plains, and one of the areas with the greatest contraction of cropland was the southeastern United States. While the intensification of cultivated systems has met the increase in food needs over the past 50 years and reduced the pressure to convert natural ecosystems into cropland, this, according to the MA, has come at the cost of greater pressure on inland water ecosystems, generally reduced biodiversity within agricultural landscapes, and higher energy inputs in the form of mechanization and the production of chemical fertilizers. Although cultivated systems provide only 16 percent of global water runoff, they tend to be close to human populations, and their nutrient and industrial water runoff affects about 5 billion people.

The absence of new suitable land for cultivation and the increased productivity of agricultural lands are reducing the need for agricultural expansion. Consequently, more of the land in cultivated systems is actually being cultivated, with increased intensity of cultivation, shorter fallows, and a shift from monocultures to polycultures. Farmers in North America and other areas increasingly are adopting appropriate soil conservation practices that reduce erosion, such as minimum tillage. Since 1950, one of the areas where cropland has stabilized is North America. The table below summarizes some current characteristics of cropland and forest ecosystems as reported by the MA, including the relative proportions of potential and actual areas in each system, indicated as the "share of area transformed."

Synergies

The MA approach attempts to look beyond the confines of particular systems and stresses a broad interconnected view of the costs and benefits of ecosystem conversions. Because of the connections among ecosystems, the degradation of one can have negative synergistic effects on others; however, the same connections can be harnessed through properly designed human interventions to produce positive synergistic effects. According to the MA, increasing food production in cultivated systems can have negative effects on biodiversity and water regulation, as well as increasing agricultural pollutants in the runoff; however, the interactions between human and natural systems can also have positive synergies. Agroforestry systems, if properly designed, can provide food and fuel, restore soils, and contribute to biodiversity conservation.

Characteristics of Forest Systems and Cultivated Systems Worldwide

System and Subsystem	Area (Million Square Kilometers)	Share of Terrestrial Surface of Earth (Percent)	Mean Net Primary Production[a] (Kilograms Carbon per Square Meter per Year)	Share of System Covered by Protected Areas[b] (Percent)	Share of Area Transformed[c] (Percent)
Forest/Woodland	**41.9**	**28.4**	**0.68**	**10**	**42**
Tropical/Subtropical	23.3	15.8	0.95	11	34
Temperate	6.2	4.2	0.45	16	67
Boreal	12.4	8.4	0.29	4	25
Cultivated	**35.3**	**23.9**	**0.52**	**6**	**47**
Pasture	0.1	0.1	0.64	4	11
Cropland	8.3	5.7	0.49	4	62
Mixed (Crop and Other)	26.9	18.2	0.60	6	43

[a]Net primary productivity (NPP) measures the rate at which carbon is fixed by plants, minus carbon lost through plant respiration.

[b]Protected areas in the International Union for the Conservation of Nature (IUCN) categories I through VI.

[c]For forest/woodland systems, the share of area transformed is the percentage difference between the forest types that could exist on the lands (forested biomes of the World Wildlife Fund's ecoregions land classification) and those that actually exist, as indicated by the Global Land Cover land classification system of the European Commission's Joint Research Center. For cultivated systems, the share of area transformed is the area indicated as such in the Global Land Cover land classification system.

Source: Millennium Ecosystem Assessment, Ecosystems and Human Wellbeing: Synthesis (Washington, DC: Island Press, 2005), web site www.millenniumassessment.org/en/Products.Synthesis.aspx.

Forest Land Remaining Forest Land: Changes in Forest Carbon Stocks

In the United States, the most significant pressures on the amount of carbon sequestered through forest lands are land management activities and the continuing effects of past changes in land use. These activities directly affect carbon flux by shifting the amount of carbon accumulated in forest ecosystems.[130] Land management activities affect both the stocks of carbon that can be stored in landbased carbon sinks, such as forests and soils, and the fluxes of carbon between landbased sinks and the atmosphere.The components or "pools" of forest carbon analyzed by the EPA for its most recent inventory include aboveground biomass, belowground biomass, dead wood, litter, soil organic carbon, harvested wood products in use, and harvested wood products in landfills. As a result of natural biogeochemical processes occurring in forests, as well as anthropogenic activities, carbon is constantly cycling through these components and between the forest and the atmosphere. The net change in overall forest carbon may not always be equal to the net flux between forests and the atmosphere, because timber harvests may not necessarily result in an instant return of carbon to the atmosphere. Timber harvesting transfers carbon from one of the five "forest carbon pools" to one of the two "wood products carbon pools." Once carbon is transferred to a product pool, it is emitted over time as carbon dioxide or methane as the product decays or is combusted. Emission rates vary significantly, depending on the type of product pool that houses the carbon.[131]

In the United States, enhanced forest management, regeneration of formerly cleared forest areas, and timber harvesting have resulted in net annual sequestration of carbon throughout the past decade. Since the 1920s, deforestation for agricultural purposes has become a nearly defunct practice. More recently, managed growth practices have become common in eastern forests, greatly increasing their biomass density over the past 50 years. In the 1970s and 1980s, federally sponsored tree planting and soil conservation programs were embraced.

These programs led to the reforestation of formerly harvested lands, improvement in timber management activities, soil erosion abatement, and the conversion of cropland to forests. Forest harvests have also affected carbon sequestration. The majority of harvested timber in the United States is used in wood products. The bulk of the discarded wood products is landfilled, and thus large quantities of the harvested carbon are relocated to longterm storage pools rather than to the atmosphere. The size of this longterm storage pool has increased over the past century.[132]

According to the EPA, carbon sequestration by U.S. forests totaled 753 $MMTCO_2e$ in 2003 (Table 34). Between 1990 and 2003, U.S. forests accounted for an average annual net sequestration of 832 $MMTCO_2e$, resulting from domestic forest growth and increases in forested land area; however, there was a decrease of approximately 21 percent in annual sequestration over the same period.[133]

The overall decline in forest carbon sequestration was driven by a 27percent reduction in the level of sequestration in the forest carbon pool (739 $MMTCO_2e$ in 1990 versus 537 $MMTCO_2e$ in 2003). The reduction in the forest carbon pool sequestration rate can be attributed primarily to a 110percent decline in the estimated level of sequestration in forest soils. Not only was there no forest soil sequestration in 2003 (as compared with 125 $MMTCO_2e$ sequestered in 1990), but forest soils became a source with average annual emissions of 12.0 $MMTCO_2e$ during the period 19992003. The EPA explains that the decrease in sequestration in this pool is derived from forest inventory data and is a direct consequence of changes in total forest area or changes in forest type.[134]

Table 34. Net Carbon Dioxide Sequestration in U.S. Forests and Harvested Wood Pools, 1990 and 19972003

Carbon Pool	1990	1997	1998	1999	2000	2001	2002	2003
Forests	**739**	**638**	**599**	**537**	**537**	**537**	**537**	**537**
Above-Ground Biomass	396	457	437	400	400	400	400	400
Below-Ground Biomass	77	89	85	78	78	78	78	78
Dead Wood	74	53	51	45	45	45	45	45
Litter	67	31	28	26	26	26	26	26
Soil Organic Carbon	125	8	-1	-12	-12	-12	-12	-12
Harvested Wood	210	213	206	215	211	214	214	216
Wood Products	48	58	52	62	59	59	59	60
Landfilled Wood	162	155	154	153	152	155	155	155
Total	**949**	**851**	**806**	**752**	**748**	**751**	**751**	**753**

Notes: The sums of the annual net stock changes in this table (shown in the "Total" row) represent estimates of the actual net flux between the total forest carbon pool and the atmosphere. Forest estimates are based on periodic measurements; harvested wood estimates are based on annual surveys and models. Totals may not equal sum of components due to independent rounding.

Source: U.S. Environmental Protection Agency, *Inventory of U.S. Greenhouse Gas Emissions and Sinks: 19902003,* EPA 430R05003 (Washington, DC, April 2005), web site http://yosemite.epa.gov/oar/globalwarming.nsf/content/ResourceCenter PublicationsGHGEmissionsUSEmissionsInventory2005.html.

The EPA points out that net forest growth and increasing forest area, particularly before 1997, contributed to rising sequestration. Since 1997, forest land area has remained relatively constant, and the increase in carbon density (per area) has resulted in net forest carbon sequestration. National estimates of forest land are obtained by summing State surveys for the conterminous United States. Because the State surveys are not completed each year, interpolation between data points is used to provide estimates for years without surveys.

Overall annual sequestration levels in harvested wood carbon stocks increased slightly between 1990 and 2003. The trend in net sequestration amounts has been generally upward, from 210 MMTCO_2e in 1990 to 216 MMTCO_2e in 2003 (Table 34). Annual sequestration levels in landfilled wood declined from 162 MMTCO_2e in 1990 to 155 MMTCO_2e in 2003, but that decline was offset by an increase in carbon sequestration in harvested wood products, from 48 MMTCO_2e in 1990 to 60 MMTCO_2e in 2003.

The EPA has estimated carbon stocks in wood products in use and in landfills from 1910 onward, based on USDA Forest Service historical data and analyses using the North American Pulp and Paper (NAPAP) model,[135] the Timber Assessment Market Model (TAMM),[136] and the Aggregate Timberland Assessment System (ATLAS) model.[137] Carbon decay in harvested wood was analyzed by the EPA for the period 1910 through 2003, using data on annual wood and paper production. The analysis included changes in carbon stocks in wood products, changes in carbon in landfills, and the amount of carbon emitted to the atmosphere (carbon dioxide and methane) both with and without energy recovery. The EPA also followed the "production approach"; that is, carbon stored in imported wood products was not counted, but carbon stored in exports was counted, including logs processed in other countries[138] (see box on page 82).

Croplands Remaining Croplands: Changes in Agricultural Soil Carbon Stocks

The amount of organic carbon in soils depends on the balance between the addition of organic material and the loss of carbon through decomposition. The quantity and quality of organic matter within soils, as well as decomposition rates, are determined by the interaction of climate, soil properties, and land use. Agricultural practices—including clearing, drainage, tillage, planting, grazing, crop residue management, fertilization, and flooding—can alter organic matter inputs and decomposition, causing a net flux of carbon to or from soils.

The IPCC methodology, which is used by the EPA to estimate the net flux from agricultural soils (Table 35), is divided into three categories of land-use and landmanagement activities: (1) agricultural land use and land management activities on mineral soils;[139] (2) agricultural land-use and landmanagement activities on organic soils;[140] and (3) liming of soils. Of the three activities, the use and management of mineral soils is estimated to be the most significant contributor to total carbon sequestration from 1990 through 2003. Sequestration in mineral soils in 2003 was estimated to be 51.7 MMTCO_2e, while emissions from organic soils and liming were estimated at 35.6 and 9.5 MMTCO_2e, respectively. Together, these three activities resulted in a net 6.6 MMTCO_2e sequestered through agricultural soils in 2003.[141]

Accounting for Harvested Wood Products in Future Greenhouse Gas Inventories

Harvested wood products (HWP) are defined as "goods manufactured or processed from wood, including lumber and panels for end uses such as housing and furniture, and paper and paperboard for uses such as packaging, printing and writing, and sanitary applications."[a]HWP are an important part of the overall carbon cycle and are thus integral to any greenhouse gas accounting system or inventory.

Preparation of the 2006 guidelines of the Intergovernmental Panel on Climate Change (IPCC) for preparing national greenhouse gas inventories under the United Nations Framewrk

Convention on Climate Change (UNFCCC)—including methods for estimating and reporting of HWP—is underway. The issue of accounting for HWP, however, is a complex one, and involves the consideration of factors such as international trade (importexport) of wood products, timing of emissions accounting, determining whether emissions include those from existing wood product pools or solely from harvesting, and establishing how complex or simple the accounting approach should be so as not to create barriers to participation.[b]

Three approaches—stockchange, production, and atmosphericflow—have been developed and debated, and all three were discussed at a UNFCCCsponsored workshop held in Norway on August 30 through September 2, 2004.[c] At the Eleventh Conference of the Parties to the UNFCCC in December 2005, the parties agreed to return to this issue at the 24th meeting of the Subsidiary Body for Scientific and Technical Advice.[d]

The first approach under consideration for HWP is the *stockchange approach*. This approach accounts for changes in carbon stock in forests in the country in which the wood is grown, deemed the producing country. Changes in the products pool are accounted for in the country where the products are used, deemed the consuming country. These stock changes are counted within national boundaries, where and when they occur.[b] Under this approach, the HWP stock change in a country may be estimated considering either transfers into and out of the HWP pool, or the difference betweenHWPcarbon stocks at two different set points in time.

The next alternative—the atmosphericflow approach—accounts for emissions or sequestration of carbon to and/or from the atmosphere within national boundaries, both where and when emissions and sequestration occur. The producing country accounts for sequestration of carbon attributed to forest growth, while the consuming country accounts for emissions of carbon to the atmosphere from oxidation of HWP.[b]Under this approach, it is the net flow of carbon dioxide from the pools to the atmosphere that would be reported as the equivalent emission, and the net flow in the opposite direction as the equivalent amount of carbon sequestration

The third approach for accounting for HWP is the production approach. While this approach also reports changes in carbon stock, it is the producing country that reports the stock changes inHWPregardless of the location of the stock (i.e., whether within country boundaries or exported).[c] This approach thus accounts for domestically produced stocks only; that is, stock changes are counted when they occur, regardless of where the stock change occurs.[b]

An additional method that was proposed by one Annex I country—the simple decay approach—effectively falls under the production approach. This method assumes that HWP remain a part of the forest in which they were produced until decomposed.[c] This approach is therefore similar to the production approach in that it also estimates the stock changes in HWP when, but not where, they occur if wood products are exported or traded. Both sequestration of carbon from the atmosphere due to forest growth and emissions resulting from harvesting are accounted for in the producing country.[b]

Settlements Remaining Settlements

Changes in Urban Tree Carbon Stocks

Urban forests make up a considerable portion of the total tree canopy cover in the United States. Urban areas, which cover 3.5 percent of the continental United States, are estimated to contain about 3.8 billion trees, accounting for approximately 3 percent of total tree cover in the United States. The EPA's carbon sequestration estimates for urban trees are derived from estimates by Nowak and Crane,[142] based on data collected throughout the 1990s and applied to the entire time series in this report. Net carbon dioxide sequestration from urban trees is estimated at 58.7 MMTCO$_2$e sequestered annually from 1990 through 2003 (Table 33).[143]

Changes in Landfilled Yard Trimming and Food Scrap Carbon Stocks

Carbon stored in landfilled yard trimmings can remain sequestered indefinitely. In the United States, yard trimmings (grass clippings, leaves, and branches) and food scraps make up a considerable portion of the municipal waste stream, and significant amounts of the yard trimmings and food scraps collected are discarded in landfills. Both the amount collected annually and the percentage that is landfilled have declined over the past decade. Net carbon dioxide sequestration from landfilled yard trimmings and food scraps has declined accordingly, from 26.0 MMTCO$_2$e in 1990 to 10.1 MMTCO$_2$e in 2003 (Table 36).

Since 1990, municipal policies limiting pickup and disposal have led to a 20percent decrease in yard trimmings collected. In addition, composting of yard trimmings in municipal facilities has increased significantly, reducing the percentage of total yard trimmings placed in landfills from 72 percent in 1990 to 34 percent in 2003. In contrast, the percentage of food scraps disposed of in landfills has decreased only slightly, from 81 percent in 1990 to 77 percent in 2003. The EPA's methodology for estimating carbon storage relies on a lifecycle analysis of greenhouse gas emissions and sinks associated with solid waste management.[144]

Land Use and International Climate Change Negotiations

In past international negotiations on climate change, the United States and many other countries have maintained that the inclusion of LULUCF activities in a binding agreement that limits greenhouse gas emissions is of the utmost importance; however, issues of whether and how terrestrial carbon sequestration could be accepted for meeting various commitments and targets have remained subjects of complex and difficult international negotiations.

Table 35. Net Carbon Dioxide Sequestration in U.S. Agricultural Soils, 1990 and 19972003

Description	1990	1997	1998	1999	2000	2001	2002	2003
Mineral Soils	52.4[a]	51.7[a]	49.5[b]	48.9[b]	50.0[b]	51.6[b]	51.9[b]	51.7[b]
Organic Soils	-34.8[a]	-35.6[a]	-35.6[b]	-35.6[b]	-35.6[b]	-35.6[b]	-35.6[b]	-35.6[b]
Liming of Soils	-9.5[a]	-8.7[a]	-9.6[a]	-9.1[a]	-8.8[a]	-9.0[a]	-10.1[a]	-9.5[b]
Total	**8.1[a]**	**7.4[a]**	**4.3[b]**	**4.3[b]**	**5.7[b]**	**7.1[b]**	**6.2[b]**	**6.6[b]**

[a]Estimates based on historical data.

[b]Estimates based on a combination of historical data and projections.

Note: Negative values indicate net emissions.

Source: U.S. Environmental Protection Agency, *Inventory of U.S. Greenhouse Gas* Emissions and Sinks: 19902003, EPA 430R05003 (Washington, DC, April 2005), web site http://yosemite.epa.gov/oar/globalwarming.nsf/content/ResourceCenter PublicationsGHGEmissionsUSEmissionsInventory2005.html.

Many of the countries involved in climate change negotiations have agreed that implementation of LULUCF activities under an international climate change agreement may be complicated by a lack of clear definitions of "reforestation" and "forest."

Further, implementation may be hindered by the lack of effective accounting rules. According to research published by the Pew Center on Global Climate Change,[145] implementation of LULUCF provisions in an international climate change agreement raises many issues, such as:

- What is a direct humaninduced activity?
- What is a forest and what is reforestation?
- How will the issues of uncertainty and verifiability be addressed?
- How will the issues of (non) permanence and leakage be addressed?
- Which activities beyond afforestation, reforestation and deforestation (ARD), if any, should be included, and what accounting rules should apply?
- Which carbon pools and which greenhouse gases should be considered?

Uncertainties related to data issues have also slowed international negotiations on climate change.

The Ninth Session of the Conference of the Parties to the UN Framework Convention on Climate Change (COP9) was held in Milan, Italy, in December 2003. The parties agreed on some of the rules for carbon sequestration projects under the Clean Development Mechanism (CDM), but the issue of how to treat the nonpermanence of carbon sinks projects remained unresolved. Delegates at COP9 decided to limit the duration of credits generated from carbon sequestration projects and addressed the topics of additionality, leakage, uncertainties, and socioeconomic and environmental impacts.[146]

A year later in Buenos Aires, Argentina, delegates at the Tenth Conference of the Parties (COP10) did address the issue of smallscale afforestation and reforestation project activities under the CDM. The following decisions were made at COP10:[147]

- Adopt simplified modalities and procedures for smallscale afforestation and reforestation project activities in the first commitment period.
- Limit the designation of smallscale afforestation and reforestation projects to those with net anthropogenic greenhouse gas removals by sinks that are less than 8,000 metric tons carbon dioxide equivalent per year. For projects that result in greenhouse gas removals of more than this quantity, the excess would be ineligible for temporary or longterm certified emissions reductions.
- Exclude funds obtained through smallscale project activities from the share of proceeds to be used to assist developing countries particularly vulnerable to the adverse impacts of climate change. Such countries shall be entitled to a reduced level of the nonreimbursable fee for requesting registration and a reduced rate of the proceeds to cover administrative expenses of the CDM.

Table 36. Net Carbon Dioxide Sequestration from Landfilled Yard Trimmings and Food Scraps, 1990 and 19972003

(Million Metric Tons Carbon Dioxide Equivalent)

Description	1990	1997	1998	1999	2000	2001	2002	2003
Yard Trimmings	**23.2**	**10.4**	**9.6**	**8.5**	**7.2**	**7.4**	**7.5**	**7.5**
Grass	2.5	0.9	0.8	0.7	0.6	0.7	0.7	0.7
Leaves	11.2	5.4	5.1	4.5	4.0	4.0	4.0	4.0
Branches	9.6	4.0	3.7	3.2	2.6	2.7	2.7	2.8
Food Scraps	**2.8**	**2.6**	**2.9**	**2.9**	**3.0**	**2.9**	**2.7**	**2.6**
Total	**26.0**	**12.9**	**12.5**	**11.4**	**10.2**	**10.3**	**10.2**	**10.1**

Note: Totals may not equal sum of components due to independent rounding.

Source: U.S. Environmental Protection Agency, Inventory of U.S. Greenhouse Gas Emissions and Sinks: 19902003, EPA 430R05003 (Washington, DC, April 2005), web site http://yosemite.epa.gov/oar/globalwarming.nsf/content/ResourceCenter PublicationsGHGEmissionsUSEmissionsInventory2005.html.

Land-use Data Issues

The EPA's most recent inventory report discusses the uncertainty inherent in the methodology used to estimate forest carbon stocks.[148] The estimates of forest carbon in live biomass, dead wood, and litter are based on USDA forest survey data for the conterminous United States, because no survey data are available for Alaska, Hawaii, and the U.S. Territories. The survey data are statistical samples designed to represent vast areas of land. The USDA mandates that the survey data be accurate to within 3 percent, at a confidence level of 67 percent.[149] An analysis of this methodology for the southeastern United States showed that the uncertainty resulted from sampling errors and not from the regression equations used to calculate tree volume (and thus carbon) from survey statistics such as tree height and diameter. The standard errors of 1 to 2 percent for volumes of growing stock in individual States are insignificant; however those for changes in the volumes of growing stock are much higher, ranging from 12 percent to as much as 139 percent.[150]

Additional uncertainty is associated with the estimates of carbon stocks in other carbon pools, which are based on extrapolations of the relationships among variables in sitespecific studies to all forest land. Such extrapolation is needed in the absence of survey data on other carbon pools.[151] The extrapolations bring in uncertainty from modeling errors and conversions between different reporting units. The effect of land-use change and forest management activities (such as harvest) on soil stocks is another large source of uncertainty, with little consensus in the literature.

Global Forest ReSources Assessment 2000

The Food and Agriculture Organization of the United Nations (FAO) is the main intergovernmental source of data on global forests. FAO's global forest assessments date back to 1948, with the most recent assessment— Global Forest Resources Assessment 2000—published in 2001. The 2000 assessment was the first to include a uniform definition of forests for all regions of the world—that is, areas with at least 10 percent of canopy cover (excluding stands of trees primarily used for agricultural production). Using this new definition, FAO estimated the world's forested area in 2000 at 3.9 billion hectares.

The 2000 assessment reported that the world's forests showed average net annual losses of 9.4 million hectares from 1990 to 2000, with annual losses of 14.6 million hectares due to deforestation and annual gains of 5.2 million hectares due to reforestation, afforestation, and the natural expansion of forests. Net losses for tropical forests were 12.3 million hectares annually, and net gains for nontropical forests were 2.9 million hectares annually [a]

The FAO Global Forest Resources Assessment 2000 draws its forest data for the United States from U.S. Forest Service periodic forest inventories, which cover all forest land in the United States for more than 70 years. The Forest Inventory and Analysis (FIA) traditionally sampled on a 5 to 10year cycle with an accuracy of ±1 percent per million hectares for forest area estimates. Since 1996, however, the FIA has involved annual sampling in many States. Currently, 46 States are sampled annually. The FAO Assessment for 2000 cites total U.S. forest area at 226 million hectares. The change in U.S. forest area from 1990 to 2000 was 0.4 million hectares per year[b]

A revised assessment is currently being prepared and will be published in early 2006. Global Forest Resources Assessment 2005 will involve more sophisticated datasets that result from satellite remote sensing [c] At the time this chapter was written, the assessment had not yet been published.

REFERENCES

Executive Summary

[1] See "Units for Measuring Greenhouse Gases" on page xi, and Intergovernmental Panel on Climate Change, *Climate Change 2001: The Scientific Basis* (Cambridge, UK: *Cambridge University Press*, 2001).

[2] Intergovernmental Panel on Climate Change, *Climate Change 2001: The Scientific Basis* (Cambridge, UK: *Cambridge University Press*, 2001).

[3] Intergovernmental Panel on Climate Change, *Climate Change 1995: The Science of Climate Change* (Cambridge, UK: *Cambridge University Press*, 1996).

[4] Energy Information Administration, *ShortTerm Energy Outlook* (Washington, DC, July 2005), Appendix A, Table A2, p. 2, web site www.eia.doe.gov/emeu/steo/pub/pdf/a2tab.pdf.

[5] Sectoral (residential, commercial, and industrial) energyrelated carbon dioxide emissions include the share of total electric power sector carbon dioxide emissions that can be attributed to each enduse sector. The share is based on the percentage of total electricity sales purchased by the sector and losses attributed to the sector. (For values used to calculate sectoral shares, see Energy Information Administration, *Monthly Energy Review,* DOE/EIA0035, Tables 2.2, 2.3, 2.4, and 2.5, web site www.eia.doe.gov/emeu/mer/consump.html.) All carbon dioxide emissions associated with industrial or commercial enterprises whose primary business is not the production of electricity are allocated to the sectors in which they occur.

[6] Unlike emissions of carbon that occur in the form of carbon dioxide, carbon is sequestered in combination with other chemicals. Therefore, carbon sequestration is estimated in carbon dioxide equivalent units, rather than simply as carbon dioxide.

[7] Data for 2004 military bunker fuels were not available at the time of publication. It should also be noted that only bunker fuels purchased in the United States are subject to adjustment.

[8] Based on an estimated GWP of 23 for methane. For an expanded discussion of global warming potentials, see Chapter 1, page 10.

[9] U.S. Environmental Protection Agency, *Inventory of U.S. Greenhouse Gas Emissions and Sinks: 19902003,* EPA 430R05003 (Washington, DC, April 2005), web site http://yosemite.epa.gov/oar/globalwarming.nsf/content/ResourceCenterPublicationsGHG EmissionsUS EmissionsInventory2005.html. Estimates for carbon sequestration in 2004 are not yet available.

[10] EIA does not include sequestration from land-use change and forestry as part of its annual emissions inventory.

[11] Energy Information Administration, *Documentation for Estimation of Greenhouse Gases in the United States 2003,* DOE/EIA0638(2003) (Washington, DC, May 2005), Chapter 8, web site http://www.eia.doe.gov/oiaf/1605/ggrpt/documentation/pdf/0638(2003).pdf.

U.S. Emissions of Greenhouse Gases: Background and Context

[1] U.S. Environmental Protection Agency, *Inventory of U.S. Greenhouse Gas Emissions and Sinks: 19902003,* EPA 430R05003 (Washington, DC, April 2005), web site http://yosemite.epa.gov/oar/globalwarming.nsf/content/ResourceCenterPublicatio nsGHGEmissionsUS EmissionsInventory2005.html.

[2] Intergovernmental Panel on Climate Change, Good Practice Guidance for Land Use, Land-use Change and Forestry (Hayama, Japan, 2003), web site www.ipccnggip.iges.or.jp/public/gpglulucf/gpglulucf.htm.

[3] The State Soil Geographic (STATSGO) database is a 1:250,000 scale generalized soils database, prepared by the National Resources Conservation Service of the U.S. Department of Agriculture. See web site www.ncgc.nrcs.usda.gov/products/datasets/statsgo.

[4] The USDA's Forest Inventory and Analysis (FIA) Program provides the information needed to assess forests in the United States. Through an annual survey, FIA reports on status and trends in forest area and location. See web site http://fia.fs.fed.us.

[5] The Forest and Rangeland Renewable Resources Planning Act of 1974 (RPA) requires the Secretary of Agriculture to conduct an assessment of the Nation's renewable resources every 10 years. In the most recent assessment, which was done in 2000, the emphasis was expanded from solely economic concerns to resource conditions, ecosystem health, and sustainability. See web site www.fs.fed.us/pl/rpa/what.htm.

[6] Energy Information Administration, *International Energy Outlook* 2005, DOE/EIA0484(2005) (Washington, DC, July 2005). The historical estimates and

projections of U.S. energyrelated carbon dioxide emissions have been revised; emissions estimates for the rest of the world have not been revised from those published in the *International Energy Outlook* 2005. Emissions of gases other than energyrelated carbon dioxide are difficult to estimate for the developing world; however, emissions related to fossil fuel consumption are likely to make up 80 to 85 percent of total greenhouse gas emissions.

[7] Intergovernmental Panel on Climate Change, Climate Change 2001*: The Scientific Basis* (Cambridge, UK: *Cambridge University Press*, 2001), pp. 8990. See also web site www.ipcc.ch.

[8] The UNFCCC, which "entered into force" in 1994, called on Annex I countries defined in the Convention, including the United States, to return their greenhouse gas emissions to 1990 levels by the year 2000. The Kyoto Protocol, adopted in December 1997, set quantified greenhouse gas emissions targets for Annex I countries for the 2008 to 2012 commitment period that are collectively about 5 percent lower than the 1990 emissions of those countries. The United States, at UNFCCC negotiations in Bonn, Germany, in July 2001, indicated that it considered the Kyoto Protocol to be flawed and stated that it had no plans to ratify the Protocol. The Kyoto Protocol entered into force in February 2005, 3 months after signatory countries accounting for 61 percent of total 1990 Annex I carbon dioxide emissions had ratified the agreement.

[9] Intergovernmental Panel on Climate Change, Climate Change 2001*: The Scientific Basis* (Cambridge, UK: *Cambridge University Press*, 2001), p. 26.

[10] Intergovernmental Panel on Climate Change, Climate Change 2001*: The Scientific Basis* (Cambridge, UK: *Cambridge University Press*, 2001), p. 10.

[11] National Research Council, Climate Change Science*, An Analysis of Some Key Questions* (Washington, DC: *National Academy Press*, 2001), Appendix A, "Letter from the White House," p. 27, web site http://nap.edu/html/climatechange/.

[12] National Research Council, Climate Change Science*, An Analysis of Some Key Questions* (Washington, DC: *National Academy Press*, 2001), p. 1, web site http://nap.edu/html/climatechange/.

[13] Intergovernmental Panel on Climate Change, Climate Change 2001*: The Scientific Basis. Summaryfor Policymakers* (Cambridge, UK: *Cambridge University Press*, 2001), p. 49.

[14] Intergovernmental Panel on Climate Change, Climate Change 2001*: The Scientific Basis. Summaryfor Policymakers* (Cambridge, UK: *Cambridge University Press*, 2001), Table 6.1, p. 358.

[15] Intergovernmental Panel on Climate Change, Climate Change 2001*: The Scientific Basis. Summaryfor Policymakers* (Cambridge, UK: *Cambridge University Press*, 2001), p. 185.

[16] Intergovernmental Panel on Climate Change, Climate Change 2001*: The Scientific Basis* (Cambridge, UK: *Cambridge University Press*, 2001), pp. 188, 191.

[17] Intergovernmental Panel on Climate Change, *Land Use, Land-use Change, and Forestry. A Special Report to the IPCC* (Cambridge, UK: *Cambridge University Press*, 2000).

[18] Intergovernmental Panel on Climate Change, Climate Change 2001*: The Scientific Basis* (Cambridge, UK: Cambridge *University Press, 2001*), p. 208.

[19] Intergovernmental Panel on Climate Change, Climate Change 2001: *The Scientific Basis* (Cambridge, UK: *Cambridge University Press*, 2001), p. 248.
[20] Intergovernmental Panel on Climate Change, Climate Change 2001: *The Scientific Basis* (Cambridge, UK: *Cambridge University Press,* 2001), Table 4.2, p. 250.
[21] Intergovernmental Panel on Climate Change, Climate Change 2001: *The Scientific Basis. Summaryfor Policymakers* (Cambridge, UK: *Cambridge University Press*, 2001), Table 6.1, p. 358.
[22] Intergovernmental Panel on Climate Change, Climate Change 2001: *The Scientific Basis* (Cambridge, UK: *Cambridge University Press*, 2001), Table 4.4, p. 252.
[23] Intergovernmental Panel on Climate Change, Climate Change 2001: *The Scientific Basis. Summaryfor Policymakers* (Cambridge, UK: *Cambridge University Press*, 2001), Table 6.1, p. 358.
[24] Intergovernmental Panel on Climate Change, Climate Change 2001: *The Scientific Basis* (Cambridge, UK: *Cambridge University Press*, 2001), Figures 4.6 and 4.7, p. 255.
[25] Intergovernmental Panel on Climate Change, Climate Change 2001: *The Scientific Basis* (Cambridge, UK: *Cambridge University Press*, 2001), p. 254; and estimates presented in this report.
[26] Intergovernmental Panel on Climate Change, Climate Change 2001: *The Scientific Basis* (Cambridge, UK: *Cambridge University Press*, 2001), p. 254.
[27] Intergovernmental Panel on Climate Change, Climate Change 2001: *The Scientific Basis* (Cambridge, UK: *Cambridge University Press*, 2001), p. 254.
[28] W.T. Sturges et al., "A Potent Greenhouse Gas Identified in the Atmosphere: SF_5CF_3," *Science,* Vol. 289 (July 28,2000), pp. 611613. [29]Intergovernmental Panel on Climate Change, Climate Change 2001: *The Scientific Basis* (Cambridge, UK: *Cambridge University Press*, 2001), p. 254.
[30] Intergovernmental Panel on Climate Change, Climate Change 2001: *The Scientific Basis. Summaryfor Policymakers* (Cambridge, UK: *Cambridge University Press*, 2001), p. 43.
[31] Intergovernmental Panel on Climate Change, Climate Change 2001: *The Scientific Basis. Summaryfor Policymakers* (Cambridge, UK: *Cambridge University Press*, 2001), p. 43.
[32] Intergovernmental Panel on Climate Change, Climate Change 2001: *The Scientific Basis. Summaryfor Policymakers* (Cambridge, UK: *Cambridge University Press*, 2001), p. 44.
[33] Intergovernmental Panel on Climate Change, Climate Change 2001: *The Scientific Basis. Summaryfor Policymakers* (Cambridge, UK: *Cambridge University Press*, 2001), p. 44.
[34] Intergovernmental Panel on Climate Change, Climate Change 2001: *The Scientific Basis* (Cambridge, UK: *Cambridge University Press*, 2001), Table 4.7(a), p. 258.
[35] Intergovernmental Panel on Climate Change, Climate Change 2001: *The Scientific Basis. Summaryfor Policymakers* (Cambridge, UK: *Cambridge University Press*, 2001), p. 45.
[36] Intergovernmental Panel on Climate Change, Climate Change 2001: *The Scientific Basis. Summaryfor Policymakers* (Cambridge, UK: *Cambridge University Press*, 2001), p. 45.

[37] Intergovernmental Panel on Climate Change, Climate Change 1994: *Radiative Forcing of Climate Change* (Cambridge, UK: *Cambridge University Press*, 1995).

[38] Intergovernmental Panel on Climate Change, Climate Change 1995: *The Science of Climate Change* (Cambridge, UK: *Cambridge University Press*, 1996).

[39] Intergovernmental Panel on Climate Change, Climate Change 2001: *The Scientific Basis* (Cambridge, UK: *Cambridge University Press,* 2001).

[40] Intergovernmental Panel on Climate Change, Climate Change 2001: *The Scientific Basis* (Cambridge, UK: *Cambridge University Press*, 2001), pp. 385386.

[41] Intergovernmental Panel on Climate Change, Climate Change 2001: *The Scientific Basis* (Cambridge, UK: *Cambridge University Press*, 2001), pp. 387390.

[42] See "White House Reviews Action on Global Climate Change," Office of the Press Secretary, The White House, updated by the Council on Environmental Quality (September 22, 2004), web site www.whitehouse.gov/ceq/globalchange.html

[43] U.S. Environmental Protection Agency, Climate Leaders Program, "Annual GHG Inventory Summary and Goal Tracking Form," web site www.epa.gov/ climateleaders/docs/summaryform.xls. For information on Climate Leaders Program reporting requirements, see web site www.epa.gov/climateleaders/ resources/reporting.html.

[44] For the agenda, a list of attendees, the presentation slides, a transcript, and an audio recording of the proceedings see web site www.pi.energy.gov/ enhancingGHGregistry.

[45] The results of these evaluations, as well as other information about the USDA workshop, are available from web site www.usda.gov/ agency/oce/gcpo/ greenhousegasreporting.htm#Febworkshop.

[46] Comments received by DOE are posted on web site https://ostiweb.osti.gov/pighg/ghgc0202.idc.

[47] The Framework Convention was "adopted" by a vote of the conference of the parties on May 9th, while the signatures and ratifications of member states flowed in over a period of years. The treaty "entered into force" in 1994. For a discussion of the development of the Convention, see D. Bodanzky, "Prologue to the Climate Convention," in I. Mintzer and J.A. Leonard (eds.), Negotiating Climate Change: *The Inside Story of the Rio Convention* (Cambridge, UK: *Cambridge University Press,* 1994), pp. 4966.

[48] The official text of the Framework Convention can be found at web site http://unfccc.int/essential_background/convention/ background/items/2853.php.

[49] The Annex I nations include Australia, Austria, Belarus, Belgium, Bulgaria, Canada, Croatia, Czech Republic, Denmark, Estonia, European Community, Finland, France, Germany, Greece, Hungary, Iceland, Ireland, Italy, Japan, Latvia, Liechtenstein, Lithuania, Luxembourg, Monaco, Netherlands, New Zealand, Norway, Poland, Portugal, Romania, Russian Federation, Slovakia, Slovenia, Spain, Sweden, Switzerland, Ukraine, United Kingdom, and the United States of America. Turkey has not ratified the Framework Convention. Turkey will be placed in a different situation from that of other Annex I parties when it becomes a Party to the Convention. Kazakhstan has announced its intention to be bound by Annex I commitments, but is not formally classified as an Annex I party. Kazakhstan was, however, considered an

Annex I party when the Kyoto Protocol entered into force in February 2005. Turkey acceded to the Convention in May 2004.

[50] The text of the Kyoto Protocol can also be found at web site www.unfccc.de/index.html.

[51] For details on the Kyoto Protocol see archived editions of this report, web site www.eia.doe.gov/oiaf/1605/1605aold.html.

[52] Earlier COP sessions are described in previous editions of this report (web site www.eia.doe.gov/oiaf/1605/1605aold.html).

[53] This summary of the G8 Summit is based on a Pew Center synopsis, available from web site www.pewclimate.org/policy_center/ international_policy/ summary_of_g8.cfm.

[54] Included as noncarbon renewable fuels are hydropower, wood, solar, and wind. Both geothermal power and waste combustion produce some carbon dioxide emissions. Woodfired generation is considered carbonneutral so long as it does not lead to deforestation.

[55] Unlike emissions of carbon that occur in the form of carbon dioxide, carbon is sequestered in combination with other chemicals. Therefore, carbon sequestration is estimated in carbon dioxide equivalent units, rather than simply as carbon dioxide.

[56] Sectoral (residential, commercial, and industrial) energyrelated carbon dioxide emissions include the share of total electric power sector carbon dioxide emissions that can be attributed to each enduse sector. The share is based on the percentage of total electricity sales purchased by the sector and losses attributed to the sector. (For values used to calculate sectoral shares, see Energy Information Administration, *Monthly Energy Review,* DOE/EIA0035, Tables 2.2, 2.3, 2.4, and 2.5, web site www.eia.doe.gov/emeu/mer/consump.html.) All carbon dioxide emissions associated with industrial or commercial enterprises whose primary business is not the production of electricity are allocated to the sectors in which they occur.

[57] Energy Information Administration, *ShortTerm Energy Outlook* (Washington, DC, July 2005), Appendix A, Table A2, p. 2, web site www.eia.doe.gov/emeu/steo/pub/pdf/a2tab.pdf.

[58] Energy Information Administration, *Annual Energy Outlook 2005,* DOE/EIA0383(2005) (Washington, DC, February 2005), Table A5, web site www.eia.doe.gov/oiaf/aeo/excel/aeo_base.xls.

[59] Data for 2004 military bunker fuels were not available at the time of publication. It should also be noted that only bunker fuels purchased in the United States are subject to adjustment.

[60] Energy Information Administration, *Documentation for Emissions of Greenhouse Gases in the United States 2003,* DOE/EIA0638(2003) (Washington, DC, January 2004), web site www.eia.doe.gov/oiaf/1605/ggrpt/documentation/ pdf/0638(2003).pdf. See also Energy Information Administration, *Documentation for Emissions of Greenhouse Gases in the United States 2004* (to be published).

[61] U.S. Department of the Interior, U.S. Geological Service, "Cement," *Mineral Commodity Summary,* web site http://minerals.usgs.gov/minerals/pubs/commodity/ cement/cemenmcs04.pdf.

[62] U.S. Environmental Protection Agency, *Inventory of U.S. Greenhouse Gas Emissions and Sinks: 19902003,* EPA 430R05003 (Washington, DC, April 2005), web site

http://yosemite.epa.gov/oar/globalwarming.nsf/content/ResourceCenterPublicationsGHGEmissions USEmissionsInventory2005.html.

[63] Based on an estimated global warming potential factor of 23 for methane. For an expanded discussion of global warming potentials, see Chapter 1, page 10.

[64] Wherever possible, estimates of methane emissions are based on measured data. In some cases, however, measured data are incomplete or unavailable. In the absence of measured data, emissions are indexed to some known activity data, such as coal production or natural gas throughput, and multiplied by an emissions factor derived from a small sample of the relevant emissions source or through laboratory experiments. For a more detailed discussion of where measured data were used and how emissions factors were developed, see Energy Information Administration, *Documentation for Emissions of Greenhouse Gases in the United States 2003,* DOE/EIA0638(2003) (Washington, DC, January2005), web site www.eia.doe.gov/oiaf/1605/ggrpt/documentation/pdf/0638(2003).pdf; and Energy Information Administration, *Documentation for Emissions of Greenhouse Gases in the United States 2004* (to be published).

[65] Energy Information Administration, *U.S. Coal Supply and Demand: 2004 Review* (Washington, DC, April 2005), web site www.eia.doe.gov/cneaf/coal/page/special/feature04.pdf.

[66] The U.S. Environmental Protection Agency (EPA) currently accounts for methane recovery from coal mines in the year the coal seam is mined through. The relationship between estimates of emissions from degasification and estimates of gas recovery is under review and maybe revised in the future.

[67] The EPA estimates that companies participating in the Natural Gas STAR program together avoided emissions of 27.0 MMTCO_2e of methane from the natural gas system in 2004 (Table 17). Program participants report annually on emissions reductions achieved through such activities as equipment replacement, enhanced inspection and maintenance, and improved operations management. Participating companies may either use their own techniques to estimate reductions achieved or employ default values developed by the EPA and the Gas Technology Institute (formerly the Gas Research Institute).

[68] For further details see Energy Information Administration, *Documentation for Emissions of Greenhouse Gases in the United States 2003,* DOE/EIA0638(2003) (Washington, DC, January 2005), web site www.eia.doe.gov/oiaf/1605/ggrpt/documentation/pdf/0638(2003).pdf; and Energy Information Administration, *Documentation for Emissions of Greenhouse Gases in the United States 2004* (to be published).

[69] U.S. Energy Information Administration, Residential Energy Consumption Survey, web site www.eia.doe.gov/emeu/recs/recs2001/ publicuse2001.html.

[70] For a more detailed discussion of the revisions in estimation methods for mobile sources, see Energy Information Administration, *Documentation for Emissions of Greenhouse Gases in the United States 2003,* DOE/EIA0638(2003) (Washington, DC, January 2005), web site www.eia.doe.gov/oiaf/1605/ggrpt/documentation/pdf/0638(2003).pdf; and Energy Information Administration, *Documentation for Emissions of Greenhouse Gases in the United States 2004* (to be published).

[71] U.S. Environmental Protection Agency, *Inventory of U.S. Greenhouse Gas Emissions and Sinks 19902003,* EPA430R05003 (Washington, DC, April 2005), web site http:/

/yosemite.epa.gov/oar/globalwarming.nsf/content/ResourceCenterPublicationsGHGEmissionsUS EmissionsInventory2005.html.

[72] Personal communication with Brian Guzzone, U.S. Environmental Protection Agency, Landfill Methane Outreach Program.

[73] U.S. Environmental Protection Agency, Landfill Methane Outreach Program, web site www.epa.gov/lmop.

[74] U.S. Environmental Protection Agency, *Inventory of U.S. Greenhouse Gas Emissions and Sinks 19902003,* EPA430R05003 (Washington, DC, April 2005), web site http://yosemite.epa.gov/oar/globalwarming.nsf/content/ResourceCenterPublicationsGHGEmissionsUS EmissionsInventory2005.html; and U.S. Energy Information Administration, Voluntary Reporting of Greenhouse Gases database, web site www.eia.doe.gov/oiaf/1605/databases.html.

[75] U.S. Environmental Protection Agency, *Inventory of U.S. Greenhouse Gas Emissions and Sinks 19902003,* EPA430R04003 (Washington, DC, April 2005), p. 200, web site http://yosemite.epa.gov/oar/globalwarming.nsf/content/ResourceCenterPublications GHGEmissions USEmissionsInventory2005.html.

[76] U.S. Department of Agriculture, National Agricultural Statistics Service, *Crop Production—Annual Summary* (Washington, DC, various years), web site http://usda.mannlib.cornell.edu/reports/nassr/field/pcpbban.

[77] American Chemistry Council (formerly the Chemical Manufacturers Association), *Guide to the Business of Chemistry,* Table 3.12, "Production of the Top 100 Chemicals" (Washington, DC, 2003).

[78] American Iron and Steel Institute, *Annual Statistical Report,* Tables 26,31, and 32 (Washington, DC, various years).

[79] U.S. Environmental Protection Agency, web site www.epa.gov/methane/intlanalyses.html.

[80] A.F. Bouwman, "Exchange of Greenhouse Gases Between Terrestrial Ecosystems and the Atmosphere," in A.F. Bouwman (ed.), *Soils and the Greenhouse Effect* (New York, NY: John Wiley and Sons, 1990).

[81] Intergovernmental Panel on Climate Change, *Greenhouse Gas Inventory Reference Manual: Revised 1996 IPCC Guidelines for National Greenhouse Gas Inventories,* Vol. 3 (Paris, France, 1997), pp. 4.874.100, web site www.ipcc.ch/pub/guide.htm.

[82] U.S. Department of Commerce, Bureau of Census, *Current Industrial Reports: Fertilizer Materials and Related Products, Fourth Quarter 2004,* MQ325B(04)4 (Washington, DC, April 2005), Table 1.

[83] M.H. Thiemens and W.C. Trogler, "Nylon Production: An Unknown Source of Atmospheric Nitrous Oxide," *Science,* Vol. 251, No. 4996 (February 1991).

[84] Radian Corporation, *Nitrous Oxide Emissions From Adipic Acid Manufacturing* (Rochester, NY, January 1992), p. 10.

[85] R.A. Reimer, R.A. Parrett, and C.S. Slaten, "Abatement of N2O Emissions Produced in Adipic Acid," in *Proceedings of the Fifth International Workshop on Nitrous Oxide Emissions* (Tsukuba, Japan, July 1992).

[86] Food and Agriculture Organization of the United Nations, statistical databases, web site http://apps.fao.org/.

[87] Biochemical oxygen demand (BOD) is a measure of the organic content of wastewater that is subject to decomposition.

[88] Preliminary data estimates received by EIA from the EPA's Office of Air and Radiation, October 2005.

[89] Preliminary data estimates received by EIA from the EPA's Office of Air and Radiation, October 2005. Note that EIA calculates emissions in carbon dioxide equivalent units using the GWP values published by the IPCC in 2001 in its Third Assessment Report, whereas the EPA uses the GWP values from the IPCC's 1996 Second Assessment Report.

[90] European Fluorocarbon Technical Committee, web site www.fluorocarbons.org/en/applications/other_app/firefighting.html.

[91] European Fluorocarbon Technical Committee, web site www.fluorocarbons.org/en/applications/insulation_foams.html.

[92] Preliminary data estimates received by EIA from the EPA's Office of Air and Radiation, October 2005.

[93] Preliminary data estimates received by EIA from the EPA's Office of Air and Radiation, October 2005.

[94] Preliminary data estimates received by EIA from the EPA's Office of Air and Radiation, October 2005.

[95] U.S. Environmental Protection Agency, *Inventory of U.S. Greenhouse Gas Emissions and Sinks 19902003,* EPA430R05003 (Washington, DC, April 2005), web site http://yosemite.epa.gov/oar/globalwarming.nsf/content/ResourceCenterPublicationsGHGEmissions USEmissionsInventory2005.html.

[96] Preliminary data estimates received by EIA from the EPA's Office of Air and Radiation, October 2005.

[97] Alternative Fluorocarbons Environmental Acceptability Study, Production, Sales and Atmospheric Release, web site www.afeas.org/2003/html/hfc134a.html.

[98] Preliminary data estimates received by EIA from the EPA's Office of Air and Radiation, October 2005.

[99] J. Ouellette, "Fluorocarbon Market Is Poised To Grow," *Chemical Market Reporter* (June 19,2000).

[100] Preliminary data estimates received by EIA from the EPA's Office of Air and Radiation, October 2005.

[101] Preliminary data estimates received by EIA from the EPA's Office of Air and Radiation, October 2005.

[102] Email correspondence with the Office of Policy, U.S. Department of Energy, October 18,2000.

[103] Preliminary data estimates received by EIA from the EPA's Office of Air and Radiation, October 2005. Note that EIA calculates emissions in carbon dioxide equivalent units using the GWP values published by the IPCC in 2001 in its Third Assessment Report, whereas the EPA uses the GWP values from the *IPCC's* 1996 Second Assessment Report.

[104] Preliminary data estimates received by EIA from the EPA's Office of Air and Radiation, October 2005.

[105] Preliminary data estimates received by EIA from the EPA's Office of Air and Radiation, October 2005.

[106] U.S. Department of the Interior, U.S. Geological Survey, *Mineral Commodity Summaries 2003,* web site http://minerals.usgs.gov/minerals/pubs/mcs/2005/ mcs2005.pdf.

[107] Preliminary data estimates received by EIA from the EPA's Office of Air and Radiation, October 2005.

[108] Intergovernmental Panel on Climate Change, Climate Change 2001*: The Scientific Basis* (Cambridge, UK: *Cambridge University Press*, 2001), p. 389.

[109] Preliminary data estimates received by EIA from the EPA's Office of Air and Radiation, October 2005.

[110] Intergovernmental Panel on Climate Change, Climate Change 2001*: The Scientific Basis* (Cambridge, UK: *Cambridge University Press*, 2001), p. 388.

[111] Preliminary data estimates received by EIA from the EPA's Office of Air and Radiation, October 2005. Note that EIA calculates emissions in carbon dioxide equivalent units using the GWP values published by the *IPCC* in 2001 in its Third Assessment Report, whereas the EPA uses the GWP values from the IPCC's 1996 Second Assessment Report.

[112] C. Boswell, "Hydrofluorocarbons Build with Transition Away from CFCs," *Chemical Market Reporter* (September 13, 1999).

[113] C. Boswell, "Hydrofluorocarbons Build with Transition Away from CFCs," *Chemical Market Reporter* (September 13, 1999).

[114] Preliminary data estimates received by EIA from the EPA's Office of Air and Radiation, October 2005. In compiling its estimates, the EPA receives data from participants in the SF6 Emission Reduction Partnership for Electric Power Systems and the SF6 Emission Reduction Partnership for the Magnesium Industry.

[115] European Fluorocarbon Technical Committee, web site www.fluorocarbons.org/ en/families/sf6/main_app.html.

[116] U.S. Environmental Protection Agency, *Inventory of U.S. Greenhouse Gas Emissions and Sinks 19902003,* EPA430R05003 (Washington, DC, April 2005),web site http:/ /yosemite.epa.gov/oar/globalwarming.nsf/content/ResourceCenterPublicationsGHG EmissionsUS EmissionsInventory2005.html.

[117] U.S. Environmental Protection Agency, Inventory of U.S. Greenhouse Gas Emissions and Sinks 19902002, EPA430R000 (Washington, DC, April 2004), web site http:/ /yosemite.epa.gov/oar/globalwarming.nsf/content/ResourceCenterPublicationsGHGEmissionsUS EmissionsInventory2004.html.

[118] European Fluorocarbon Technical Committee, web site www.fluorocarbons.org/ en/families/sf6/main_app.html#c.

[119] Historically, emissions of SF6 from the aluminum industry have been omitted from global estimates, because any emissions are expected to be insignificant. The EPA does not estimate emissions from this source due to uncertainties about the quantities used and the amounts destroyed in the applications.

[120] The net numerical difference, or "flux," between carbon sequestration and carbon release due to natural factors can be viewed as a measure of the relative contribution of biomass to the carbon cycle.

[121] Intergovernmental Panel on Climate Change, *Greenhouse Gas Inventory* Reference Manual: Revised 1996 *IPCC Guidelines for National Greenhouse Gas Inventories,*

Vol. 3 (Paris, France, 1997), web site www.ipccnggip.iges.or.jp/public/gl/invs1.htm.

[122] U.S. Environmental Protection Agency, *Inventory of U.S. Greenhouse Gas Emissions and Sinks:* 19902003, EPA 430R05003 (Washington, DC, April 2005), web site http://yosemite.epa.gov/oar/globalwarming.nsf/content/ResourceCenterPublicationsGHGEmissions USEmissionsInventory2005.html.

[123] EIA does not include sequestration from land-use change and forestry as part of its annual estimate of emissions of greenhouse gases in the United States. Note that land use refers to maintaining land within a particular category of use, such as forests remaining forests, whereas land-use change refers to changing from one land-use type to another, as when forest is converted to grasslands, or wetlands drained to create more land for agriculture.

[124] Intergovernmental Panel on Climate Change, *Good Practice Guidancefor Land Use, Land-use Change and Forestry* (Hayama, Japan, 2003), web site www.ipccnggip.iges.or.jp/public/gpglulucf/gpglulucf.htm.

[125] Intergovernmental Panel on Climate Change, *Greenhouse Gas Inventory Reference Manual: Revised 1996 IPCC Guidelines for National Greenhouse Gas Inventories,* Vol. 3 (Paris, France, 1997), web site www.ipccnggip.iges.or.jp/public/gl/invs1.htm.

[126] The State Soil Geographic (STATSGO) database is a 1:250,000 scale generalized soils database, prepared by the National Resources Conservation Service of the U.S. Department of Agriculture. See web site www.ncgc.nrcs.usda.gov/products/datasets/statsgo.

[127] The USDA's Forest Inventory and Analysis (FIA) Program provides the information needed to assess forests in the United States. Through an annual survey, FIA reports on status and trends in forest area and location. See web site http://fia.fs.fed.us.

[128] The Forest and Rangeland Renewable Resources Planning Act of 1974 (RPA) requires the Secretary of Agriculture to conduct an assessment of the Nation's renewable resources every 10 years. In the most recent assessment, which was done in 2000, the emphasis was expanded from solely economic concerns to resource conditions, ecosystem health, and sustainability. See web site www.fs.fed.us/ pl/rpa/what.htm.

[129] U.S. Environmental Protection Agency, *Inventory of U.S. Greenhouse Gas Emissions and Sinks: 19902003,* EPA 430R05003 (Washington, DC, April 2005), web site http://yosemite.epa.gov/oar/globalwarming.nsf/content/ResourceCenterPublications GHGEmissions USEmissionsInventory2005.html.

[130] U.S. Environmental Protection Agency, *Inventory of U.S. Greenhouse Gas Emissions and Sinks: 19902003,* EPA 430R05003 (Washington, DC, April 2005), web site http://yosemite.epa.gov/oar/globalwarming.nsf/content/ResourceCenterPublications GHGEmissions USEmissionsInventory2005.html.

[131] U.S. Environmental Protection Agency, *Inventory of U.S. Greenhouse Gas Emissions and Sinks 19902003,* EPA430R05003 (Washington, DC, April 2005), p. 233, web site http://yosemite.epa.gov/oar/globalwarming.nsf/content/ResourceCenter PublicationsGHG EmissionsUSEmissionsInventory2005.html.

[132] U.S. Environmental Protection Agency, *Inventory of U.S. Greenhouse Gas Emissions and Sinks* 19902003, EPA430R05003 (Washington, DC, April 2005), p. 233, web

site http://yosemite.epa.gov/oar/globalwarming.nsf/content/ResourceCenter PublicationsGHG Emissions USEmissionsInventory2005.html.

[133] U.S. Environmental Protection Agency, *Inventory of U.S. Greenhouse Gas Emissions and Sinks 19902003,* EPA430R05003 (Washington, DC, April 2005), p. 233, web site http://yosemite.epa.gov/oar/globalwarming.nsf/content/ResourceCenter PublicationsGHG Emissions USEmissionsInventory2005.html.

[134] U.S. Environmental Protection Agency, *Inventory of U.S. Greenhouse Gas Emissions and Sinks* 19902003, EPA430R05003 (Washington, DC, April 2005), p. 233, web site http://yosemite.epa.gov/oar/globalwarming.nsf/content/ResourceCenterPublicationsGH G Emissions USEmissionsInventory2005.html.

[135] P.J. Ince, *Recycling and LongRange Timber Outlook,* USDA Forest Service General Technical Report RM242 (Fort Collins, CO, February 1994).

[136] U.S. Department of Agriculture, Forest Service, *An Analysis of the Timber Situation in the United States: 1952 to 2050,* General Technical Report PNWGTR560 (Portland, OR, February 2003).

[137] J.R. Mills and J.C. Kincaid, *The Aggregate Timberland Assessment System–ATLAS: A Comprehensive Timber Projection Model,* USDA Forest Service General Technical Report PNW281 (Portland, OR, June 1992).

[138] U.S. Environmental Protection Agency, *Inventory of U.S. Greenhouse Gas Emissions and Sinks 19902003,* EPA430R05003 (Washington, DC, April 2004), p. 210, web site http://yosemite.epa.gov/oar/globalwarming.nsf/content/ResourceCenterPublicationsGH G Emissions USEmissionsInventory2005.html.

[139] Mineral soils are soils consisting predominantly of, and having their properties determined predominantly by, mineral matter. They usually contain less than 200 grams of organic carbon per kilogram of soil (less than 120 to 180 grams per kilogram if saturated with water) but may contain an organic surface layer up to 30 centimeters thick.

[140] Organic soils are soils that, when saturated with water, have 174 grams or more of organic carbon per kilogram of soil if the mineral fraction has 500 grams per kilogram or more of clay, or 116 grams per kilogram organic carbon if the mineral fraction has no clay, or has proportional intermediate contents. If the soil is never saturated with water, organic soils have 203 or more grams per kilogram of organic carbon.

[141] U.S. Environmental Protection Agency, *Inventory of U.S. Greenhouse Gas Emissions and Sinks 19902003,* EPA430R05003 (Washington, DC, April 2005), web site http://yosemite.epa.gov/oar/globalwarming.nsf/content/ResourceCenterPublicatio nsGHGEmissions USEmissionsInventory2005.html.

[142] D.J. Nowak and D.E. Crane, "Carbon Storage and Sequestration by Urban Trees in the United States," *Environmental Pollution,* Vol. 116, No. 3 (2001), pp. 381389.

[143] U.S. Environmental Protection Agency, *Inventory of U.S. Greenhouse Gas Emissions and Sinks 19902003,* EPA430R05003 (Washington, DC, April 2004), web site http://yosemite.epa.gov/oar/globalwarming.nsf/content/ResourceCenterPublicationsGH GEmissions USEmissionsInventory2005.html.

[144] U.S. Environmental Protection Agency, *Solid Waste Management and Greenhouse Gases: A LifeCycle Assessment of Emissions and Sinks,* 2nd Edition, EPA530R02006 (Washington, DC, May 2002), web site www.epa.gov/epaoswer/nonhw/ muncpl/ghg/ghg.htm.

[145] G. Marland and B. Schlamadinger, *Land Use and Global Climate Change: Forests, Land Management, and the Kyoto Protocol* (Arlington, VA: Pew Center on Global Climate Change, June 2000), p.5, web site www.pewclimate.org/docUploads/land_use.pdf.

[146] Pew Center on Global Climate Change, "Ninth Session of the Conference of the Parties to the UN Framework Convention on Climate Change" (Milan, Italy, December 112, 2003), web site www.pewclimate.org/what_s_being_done/in_the_world/cop9/index.cfm.

[147] International Institute for Sustainable Development, "Summary of the Tenth Conference of the Parties to the UN Framework Convention on Climate Change: 618 December 2004," *Earth Negotiations Bulletin,* Vol. 12, No. 260 (December 20, 2004), web site www.iisd.ca/vol12/enb12260e.html.

[148] U.S. Environmental Protection Agency, *Inventory of U.S. Greenhouse Gas Emissions and Sinks* 19902003, EPA430R05003 (Washington, DC, April 2005), web site http://yosemite.epa.gov/oar/globalwarming.nsf/content/ResourceCenter PublicationsGHGEmissions USEmissionsInventory2005.html.

[149] That is, at least 67 percent of the samples are within 3 percent of the actual forested areas.

[150] The larger errors were found to be attributable to small actual changes in volumes of growing stock, which when over or underestimated contributed disproportionately to the standard errors for total changes in the volume of growing stock.

[151] Thus, sitespecific relationships among variables are used to create models or regression equations, which are then applied to large forested areas.

Related Links

Australian Greenhouse Office. www.greenhouse.gov.au

Carbon Dioxide Information Analysis Center. http://cdiac.esd.ornl.gov

Centre for the Analysis and Dissemination of Demonstrated Energy Technologies. www.caddetee.org

Center for Renewable Energy and Sustainable Technology. http://solstice.crest.org

Climate Ark. www.climateark.org

Energy Information Administration. www.eia.doe.gov

Energy Information Administration, Greenhouse Gas Emissions and Climate Change Publications. www.eia.doe.gov/environment.html

Global Change Data & Information System. http://globalchange.gov

Global Environment Facility. www.gefweb.org

Global Climate Change from the Government of Canada.www.climatechange.gc.ca

Intergovernmental Panel on Climate Change. www.ipcc.ch

International Energy Agency. www.iea.org

International Institute for Sustainable Development. www.iisd.org

National Institute for Global Environment Change. http://nigec.ucdavis.edu

Natural Resources Defense Council, Global Warming. www.nrdc.org/globalWarming

Organization for Economic Cooperation and Development. www.oecd.org/env/

Pew Center on Climate Change. www.pewclimate.org

Resources for the Future. www.rff.org

Weathervane, Digital Forum on Global Climate Policy. www.weathervane.rff.org

Union of Concerned Scientists. www.ucsusa.org

United Nations Development Programme. www.undp.org

United Nations Environment Programme. www.unep.ch

United Nations Environment Programme, Ozone Secretariat (Montreal Protocol). www.unep.ch/ozone/home.htm

United Nations Food and Agriculture Program. www.fao.org

United Nations Framework Convention on Climate Change. www.unfccc.de

U.S. Agency for International Development. www.usaid.gov

U.S. Department of Agriculture, Global Change Program Office. www.usda.gov/oce/gcpo/

U.S. Department of Energy, Climate Vision Program. www.climatevision.gov

U.S. Department of State, Climate Change Homepage. www.state.gov/g/oes/climate/

U.S. Environmental Protection Agency, Global Warming Homepage. www.epa.gov/globalwarming/

U.S. Environmental Protection Agency, Methane to Markets Partnership. www.epa.gov/methanetomarkets/index.htm

U.S. Geological Survey, Global Change Research. http://geochange.er.usgs.gov

U.S. Global Change Research Office. http://gcrio.gcrio.org

U.S. Global Change Research Program. www.usgcrp.gov

U.S. Initiative on Joint Implementation. www.gcrio.org/usiji/

U.S. National Aeronautics and Space Administration (NASA), Global Change Master Directory. http://gcmd.gsfc.nasa.gov

U.S. National Oceanographic and Atmospheric Administration (NOAA), Office of Global Programs. www.ogp.noaa.gov

U.S. White House, Global Climate Change Policy Book. www.whitehouse.gov/news/releases/2002/02/ climatechange.html

U.S. White House, Policies in Focus, Environment. www.whitehouse.gov/infocus/environment/

World Bank Climate Change Web Site. www.worldbank.org/climatechange

World Health Organization, Protection of the Human Environment. www.who.int/phe/en/

World Meteorological Organization. www.wmo.ch

World Resources Institute. www.wri.org

Worldwatch Institute. www.worldwatch.org

World Wildlife Fund, Climate Change Campaign. www.panda.org/about_wwf/what_we_do/climate_change/index.cfm

GLOSSARY

Acid stabilization: A circumstance where the pH of the waste mixture in an animal manure management system is maintained near 7.0, optimal conditions for methane production.

Aerobic bacteria: Microorganisms living, active, or occurring only in the presence of oxygen.

Aerobic decomposition: The breakdown of a molecule into simpler molecules or atoms by microorganisms under favorable conditions of oxygenation.

Aerosols: Airborne particles.

Afforestation: Planting of new forests on lands that have not been recently forested.

Agglomeration: The clustering of disparate elements.

Airshed: An area or region defined by settlement patterns or geology that results in discrete atmospheric conditions.

Albedo: The fraction of incident light or electromagnetic radiation that is reflected by a surface or body. See *Planetary albedo.*

Anaerobes: Organisms that live and are active only in the absence of oxygen.

Anaerobic bacteria: Microorganisms living, active, or occurring only in the absence of oxygen.

Anaerobic decomposition: The breakdown of molecules into simpler molecules or atoms by microorganisms that can survive in the partial or complete absence of oxygen.

Anaerobic lagoon: A liquidbased manure management system, characterized by waste residing in water to a depth of at least six feet for a period ranging between 30 and 200 days.

Anode: A positive electrode, as in a battery, radio tube, etc.

Anthracite: The highest rank of coal; used primarily for residential and commercial space heating. It is a hard, brittle, and black lustrous coal, often referred to as hard coal, containing a high percentage of fixed carbon and a low percentage of volatile matter. The moisture content of freshmined anthracite generally is less than 15 percent. The heat content of anthracite ranges from 22 to 28 million Btu per ton on a moist, mineralmatterfree basis. The heat content of anthracite coal consumed in the United States averages 25 million Btu per ton, on the asreceived basis (i.e., containing both inherent moisture and mineral matter). Note: Since the 1980's, anthracite refuse or mine waste has been used for steam electric power generation. This fuel typically has a heat content of 15 million Btu per ton or less.

Anthropogenic: Made or generated by a human or caused by human activity. The term is used in the context of global climate change to refer to gaseous emissions that are the result of human activities, as well as other potentially climatealtering activities, such as deforestation.

API Gravity: American Petroleum Institute measure of specific gravity of crude oil or condensate in degrees. An arbitrary scale expressing the gravity or density of liquid petroleum products. The measuring scale is calibrated in terms of degrees API; it is calculated as follows: Degrees API = (141.5/sp.gr.60 deg.F/60 deg.F) 131.5.

Asphalt: A dark browntoblack cementlike material obtained by petroleum processing and containing bitumens as the predominant component; used primarily for road construction. It includes crude asphalt as well as the following finished products: cements, fluxes, the asphalt content of emulsions (exclusive of water), and petroleum distillates blended with asphalt to make cutback asphalts. Note: The conversion factor for asphalt is 5.5 barrels per short ton.

Associated natural gas: See *Associateddissolved natural gas* and *Natural gas.*

Associateddissolved natural gas: Natural gas that occurs in crude oil reservoirs either as free gas (associated) or as gas in solution with crude oil (dissolved gas). See *Natural gas.*

Aviation gasoline (finished): A complex mixture of relatively volatile hydrocarbons with or without small quantities of additives, blended to form a fuel suitable for use in aviation reciprocating engines. Fuel specifications are provided in ASTM Specification D 910 and Military Specification MILG5572. Note: Data on blending components are not counted in data on finished aviation gasoline.

Balancing item: Represents differences between the sum of the components of natural gas supply and the sum of the components of natural gas disposition. These differences may be due to quantities lost or to the effects of data reporting problems. Reporting problems include differences due to the net result of conversions of flow data metered at varying temperature and pressure bases and converted to a standard temperature and pressure base; the effect of variations in company accounting and billing practices; differences between billing cycle and calendar period time frames; and imbalances resulting from the merger of data reporting systems that vary in scope, format, definitions, and type of respondents.

Biofuels: Liquid fuels and blending components produced from biomass (plant) feedstocks, used primarily for transportation.

Biogas: The gas produced from the anaerobic decomposition of organic material in a landfill.

Biogenic: Produced by the actions of living organisms.

Biomass: Organic nonfossil material of biological origin constituting a renewable energy source.

Biosphere: The portion of the Earth and its atmosphere that can support life. The part of the global carbon cycle that includes living organisms and biogenic organic matter.

Bituminous coal: A dense coal, usually black, sometimes dark brown, often with welldefined bands of bright and dull material, used primarily as fuel in steamelectric power generation, with substantial quantities also used for heat and power applications in manufacturing and to make coke. Bituminous coal is the most abundant coal in active U.S. mining regions. Its moisture content usually is less than 20 percent. The heat content of bituminous coal ranges from 21 to 30 million Btu per ton on a moist, mineralmatterfree basis. The heat content of bituminous coal consumed in the United States averages 24 million Btu per ton, on the asreceived basis (i.e., containing both inherent moisture and mineral matter).

BOD_5: The biochemical oxygen demand of wastewater during decomposition occurring over a 5day period. A measure of the organic content of wastewater.

Bromofluorocarbons (halons): Inert, nontoxic chemicals that have at least one bromine atom in their chemical makeup. They evaporate without leaving a residue and are used in fire extinguishing systems, especially for large computer installations.

Bunker fuel: Fuel supplied to ships and aircraft, both domestic and foreign, consisting primarily of residual and distillate fuel oil for ships and kerosenebased jet fuel for aircraft. The term "international bunker fuels" is used to denote the consumption of fuel for international transport activities. *Note:* For the purposes of greenhouse gas emissions inventories, data on emissions from combustion of international bunker fuels are subtracted from national emissions totals. Historically, bunker fuels have meant only ship fuel. See *Vessel bunkering.*

Calcination: A process in which a material is heated to a high temperature without fusing, so that hydrates, carbonates, or other compounds are decomposed and the volatile material is expelled.

Calcium sulfate: A white crystalline salt, insoluble in water. Used in Keene's cement, in pigments, as a paper filler, and as a drying agent.

Calcium sulfite: A white powder, soluble in dilute sulfuric acid. Used in the sulfite process for the manufacture of wood pulp.

Capital stock: Property, plant and equipment used in the production, processing and distribution of energy resources.

Carbon black: An amorphous form of carbon, produced commercially by thermal or oxidative decomposition of hydrocarbons and used principally in rubber goods, pigments, and printer's ink.

Carbon budget: Carbon budget: The balance of the exchanges (incomes and losses) of carbon between carbon sinks (e.g., atmosphere and biosphere) in the carbon cycle. See *Carbon cycle* and *Carbon sink.*

Carbon cycle: All carbon sinks and exchanges of carbon from one sink to another by various chemical, physical, geological, and biological processes. See *Carbon sink* and *Carbon budget.*

Carbon dioxide (CO_2): A colorless, odorless, nonpoisonous gas that is a normal part of Earth's atmosphere. Carbon dioxide is a product of fossilfuel combustion as well as other processes. It is considered a greenhouse gas as it traps heat (infrared energy) radiated by the Earth into the atmosphere and thereby contributes to the potential for global warming. The global warming potential (GWP) of other greenhouse gases is measured in relation to that of carbon dioxide, which by international scientific convention is assigned a value of one (1). See *Global warming potential (GWP)* and *Greenhouse gases.*

Carbon dioxide equivalent: The amount of carbon dioxide by weight emitted into the atmosphere that would produce the same estimated radiative forcing as a given weight of another radiatively active gas. Carbon dioxide equivalents are computed by multiplying the weight of the gas being measured (for example, methane) by its estimated global warming potential (which is 21 for methane). "Carbon equivalent units" are defined as carbon dioxide equivalents multiplied by the carbon content of carbon dioxide (i.e., 12/44).

Carbon flux: See *Carbon budget.*

Carbon intensity: The amount of carbon by weight emitted per unit of energy consumed. A common measure of carbon intensity is weight of carbon per British thermal unit (Btu) of energy. When there is only one fossil fuel under consideration, the carbon intensity and the emissions coefficient are identical. When there are several fuels, carbon intensity is based on their combined emissions coefficients weighted by their energy consumption levels. See *Emissions coefficient* and *Carbon output rate.*

Carbon output rate: The amount of carbon by weight per kilowatthour of electricity produced.

Carbon sequestration: The fixation of atmospheric carbon dioxide in a carbon sink through biological or physical processes.

Carbon sink: A reservoir that absorbs or takes up released carbon from another part of the carbon cycle. The four sinks, which are regions of the Earth within which carbon behaves in a systematic manner, are the atmosphere, terrestrial biosphere (usually including freshwater systems), oceans, and sediments (including fossil fuels).

Catalytic converter: A device containing a catalyst for converting automobile exhaust into mostly harmless products.

Catalytic hydrocracking: A refining process that uses hydrogen and catalysts with relatively low temperatures and high pressures for converting middle boiling or residual material to high octane gasoline, reformer charge stock, jet fuel, and/or high grade fuel oil. The process uses one or more catalysts, depending on product output, and can handle high sulfur feedstocks without prior desulfurization.

Cesspool: An underground reservoir for liquid waste, typically household sewage.

Chlorofluorocarbon (CFC): Any of various compounds consisting of carbon, hydrogen, chlorine, and fluorine used as refrigerants. CFCs are now thought to be harmful to the earth's atmosphere.

Clean Development Mechanism (CDM): A Kyoto Protocol program that enables industrialized countries to finance emissionsavoiding projects in developing countries and receive credit for reductions achieved against their own emissions limitation targets. See *Kyoto Protocol.*

Climate: The average course or condition of the weather over a period of years as exhibited by temperature, humidity, wind velocity, and precipitation.

Climate change: A term used to refer to all forms of climatic inconsistency, but especially to significant change from one prevailing climatic condition to another. In some cases, "climate change" has been used synonymously with the term "global warming"; scientists, however, tend to use the term in a wider sense inclusive of natural changes in climate, including climatic cooling.

Clinker: Powdered cement, produced by heating a properly proportioned mixture of finely ground raw materials (calcium carbonate, silica, alumina, and iron oxide) in a kiln to a temperature of about 2,700°F.

Cloud condensation nuclei: Aerosol particles that provide a platform for the condensation of water vapor, resulting in clouds with higher droplet concentrations and increased albedo.

Coal coke: See *Coke (coal).*

Coalbed methane: Methane is generated during coal formation and is contained in the coal microstructure. Typical recovery entails pumping water out of the coal to allow the gas to escape. Methane is the principal component of natural gas. Coalbed methane can be added to natural gas pipelines without any special treatment.

Coke (coal): A solid carbonaceous residue derived from lowash, lowsulfur bituminous coal from which the volatile constituents are driven off by baking in an oven at temperatures as high as 2,000 degrees Fahrenheit so that the fixed carbon and residual ash are fused together. Coke is used as a fuel and as a reducing agent in smelting iron ore in a blast furnace. Coke from coal is grey, hard, and porous and has a heating value of 24.8 million Btu per ton.

Coke (petroleum): A residue high in carbon content and low in hydrogen that is the final product of thermal decomposition in the condensation process in cracking. This product is reported as marketable coke or catalyst coke. The conversion is 5 barrels (of 42 U.S. gallons each) per short ton. Coke from petroleum has a heating value of 6.024 million Btu per barrel.

Combustion: Chemical oxidation accompanied by the generation of light and heat.

Combustion chamber: An enclosed vessel in which chemical oxidation of fuel occurs.

Conference of the Parties (COP): The collection of nations that have ratified the Framework Convention on Climate Change (FCCC). The primary role of the COP is to keep implementation of the FCCC under review and make the decisions necessary for its effective implementation. See *Framework Convention on Climate Change (FCCC).*

Cracking: The refining process of breaking down the larger, heavier, and more complex hydrocarbon molecules into simpler and lighter molecules.

Criteria pollutant: A pollutant determined to be hazardous to human health and regulated under EPA's National Ambient Air Quality Standards. The 1970 amendments to the Clean Air Act require EPA to describe the health and welfare impacts of a pollutant as the "criteria" for inclusion in the regulatory regime.

Crop residue: Organic residue remaining after the harvesting and processing of a crop.

Cultivar: A horticulturally or agriculturally derived variety of a plant.

Deforestation: The net removal of trees from forested land.

Degasification system: The methods employed for removing methane from a coal seam that could not otherwise be removed by standard ventilation fans and thus would pose a substantial hazard to coal miners. These systems may be used prior to mining or during mining activities.

Degradable organic carbon: The portion of organic carbon present in such solid waste as paper, food waste, and yard waste that is susceptible to biochemical decomposition.

Desulfurization: The removal of sulfur, as from molten metals, petroleum oil, or flue gases.

Diffusive transport: The process by which particles of liquids or gases move from an area of higher concentration to an area of lower concentration.

Distillate fuel: A general classification for one of the petroleum fractions produced in conventional distillation operations. It includes diesel fuels and fuel oils. Products known as No. 1, No.2, and No.4 diesel fuel are used in onhighway diesel engines, such as those in trucks and automobiles, as well as offhighway engines, such as those in railroad locomotives and agricultural machinery. Products known as No. 1, No. 2, and No. 4 fuel oils are used primarily for space heating and electric power generation.

Ef flux: An outward flow.

Electrical generating capacity: The fullload continuous power rating of electrical generating facilities, generators, prime movers, or other electric equipment (individually or collectively).

EMCON Methane Generation Model: A model for estimating the production of methane from municipal solid waste landfills.

Emissions: Anthropogenic releases of gases to the atmosphere. In the context of global climate change, they consist of radiatively important greenhouse gases (e.g., the release of carbon dioxide during fuel combustion).

Emissions coefficient: A unique value for scaling emissions to activity data in terms of a standard rate of emissions per unit of activity (e.g., pounds of carbon dioxide emitted per Btu of fossil fuel consumed).

Enteric fermentation: A digestive process by which carbohydrates are broken down by microorganisms into simple molecules for absorption into the bloodstream of an animal.

Eructation: An act or instance of belching.

ETBE (ethyl tertiary butyl ether): $(CH_3)_3COC_2H$: An oxygenate blend stock formed by the catalytic etherification of isobutylene with ethanol.

Ethylene: An olefinic hydrocarbon recovered from refinery processes or petrochemical processes. Ethylene is used as a petrochemical feedstock for numerous chemical applications and the production of consumer goods.

Ethylene dichloride: A colorless, oily liquid used as a solvent and fumigant for organic synthesis, and for ore flotation.

Facultative bacteria: Bacteria that grow equally well under aerobic and anaerobic conditions.

Flange: A rib or a rim for strength, for guiding, or for attachment to another object (e.g., on a pipe).

Flared: Gas disposed of by burning in flares usually at the production sites or at gas processing plants.

Flatus: Gas generated in the intestines or the stomach of an animal.

Flue gas desulfurization: Equipment used to remove sulfur oxides from the combustion gases of a boiler plant before discharge to the atmosphere. Also referred to as scrubbers. Chemicals such as lime are used as scrubbing media.

Fluidizedbed combustion: A method of burning particulate fuel, such as coal, in which the amount of air required for combustion far exceeds that found in conventional burners. The fuel particles are continually fed into a bed of mineral ash in the proportions of 1 part fuel to 200 parts ash, while a flow of air passes up through the bed, causing it to act like a turbulent fluid.

Flux material: A substance used to promote fusion, e.g., of metals or minerals.

Fodder: Coarse food for domestic livestock. **Forestomach:** See *Rumen*.

Fossil fuel: An energy source formed in the earths crust from decayed organic material. The common fossil fuels are petroleum, coal, and natural gas.

Framework Convention on Climate Change (FCCC): An agreement opened for signature at the "Earth Summit" in Rio de Janeiro, Brazil, on June 4, 1992, which has the goal of stabilizing greenhouse gas concentrations in the atmosphere at a level that would prevent significant anthropogenically forced climate change. See *Climate change.*

Fuel cycle: The entire set of sequential processes or stages involved in the utilization of fuel, including extraction, transformation, transportation, and combustion. Emissions generally occur at each stage of the fuel cycle.

Fugitive emissions: Unintended leaks of gas from the processing, transmission, and/or transportation of fossil fuels.

Gasification: A method for converting coal, petroleum, biomass, wastes, or other carboncontaining materials into a gas that can be burned to generate power or processed into chemicals and fuels.

Gate station: Location where the pressure of natural gas being transferred from the transmission system to the distribution system is lowered for transport through small diameter, low pressure pipelines.

Geothermal: Pertaining to heat within the Earth. **Global climate change:** See *Climate change.*

Global warming: An increase in the near surface temperature of the Earth. Global warming has occurred in the distant past as the result of natural influences, but the term is today most often used to refer to the warming that some scientists predict will occur as a result of increased anthropogenic emissions of greenhouse gases. See *Climate change.*

Global warming potential (GWP): An index used to compare the relative radiative forcing of different gases without directly calculating the changes in atmospheric concentrations. GWPs are calculated as the ratio of the radiative forcing that would result from the emission of one kilogram of a greenhouse gas to that from the emission of one kilogram of carbon dioxide over a fixed period of time, such as 100 years.

Greenhouse effect: The result of water vapor, carbon dioxide, and other atmospheric gases trapping radiant (infrared) energy, thereby keeping the earth's surface warmer than it would otherwise be. Greenhouse gases within the lower levels of the atmosphere trap this radiation, which would otherwise escape into space, and subsequent reradiation of some of this energy back to the Earth maintains higher surface temperatures than would occur if the gases were absent. See *Greenhouse gases.*

Greenhouse gases: Those gases, such as water vapor, carbon dioxide, nitrous oxide, methane, hydrofluorocarbons (HFCs), perfluorocarbons (PFCs) and sulfur hexafluoride, that are transparent to solar (shortwave) radiation but opaque to longwave (infrared) radiation, thus preventing longwave radiant energy from leaving the Earth's atmosphere. The net effect is a trapping of absorbed radiation and a tendency to warm the planet's surface.

Gross gas withdrawal: The fullvolume of compounds extracted at the wellhead, including nonhydrocarbon gases and natural gas plant liquids.

Gypsum: Calcium sulfate dihydrate ($CaSO_4 \cdot 2H_2O$), a sludge constituent from the conventional lime scrubber process, obtained as a byproduct of the dewatering operation and sold for commercial use.

Halogenated substances: A volatile compound containing halogens, such as chlorine, fluorine or bromine.

Halons: See *Bromofluorocarbons.*

Heating degreedays (HDD): A measure of how cold a location is over a period of time relative to a base temperature, most commonly specified as 65 degrees Fahrenheit. The measure is computed for each day by subtracting the average of the day's high and low temperatures from the base temperature (65 degrees), with negative values set equal to zero. Each day's heating degreedays are summed to create a heating degreeday measure for a specified reference period. Heating degreedays are used in energy analysis as an indicator of space heating energy requirements or use.

Herbivore: A planteating animal.

Hydrocarbon: An organic chemical compound of hydrogen and carbon in either gaseous, liquid, or solid phase. The molecular structure of hydrocarbon compounds varies from the simple (e.g., methane, a constituent of natural gas) to the very heavy and very complex.

Hydrochlorofluorocarbons (HCFCs): Chemicals composed of one or more carbon atoms and varying numbers of hydrogen, chlorine, and fluorine atoms.

Hydrofluorocarbons (HFCs): A group of manmade chemicals composed of one or two carbon atoms and varying numbers of hydrogen and fluorine atoms. Most HFCs have 100year Global Warming Potentials in the thousands.

Hydroxyl radical (OH): An important chemical scavenger of many trace gases in the atmosphere that are greenhouse gases. Atmospheric concentrations of OH affect the atmospheric lifetimes of greenhouse gases, their abundance, and, ultimately, the effect they have on climate.

Intergovernmental Panel on Climate Change (IPCC): A panel established jointly in 1988 by the World Meteorological Organization and the United Nations Environment Program to assess the scientific information relating to climate change and to formulate realistic response strategies.

International bunker fuels: See *Bunker fuels.*

Jet fuel: A refined petroleum product used in jet aircraft engines. It includes kerosenetype jet fuel and naphthatype jet fuel.

Joint Implementation (JI): Agreements made between two or more nations under the auspices of the Framework Convention on Climate Change (FCCC) whereby a developed country can receive "emissions reduction units" when it helps to finance projects that reduce net emissions in another developed country (including countries with economies in transition).

Kerosene: A light petroleum distillate that is used in space heaters, cook stoves, and water heaters and is suitable for use as a light source when burned in wickfed lamps. Kerosene has a maximum distillation temperature of 400 degrees Fahrenheit at the 10percent recovery point, a final boiling point of 572 degrees Fahrenheit, and a minimum flash point of 100 degrees Fahrenheit. Included are No. 1K and No. 2K, the two grades recognized by ASTM Specification D 3699 as well as all other grades of kerosene called range or stove oil, which have properties similar to those of No. 1 fuel oil. See *Kerosenetype jet fuel.*

Kerosenetype jet fuel: A kerosenebased product having a maximum distillation temperature of 400 degrees Fahrenheit at the 10percent recovery point and a final maximum boiling point of 572 degrees Fahrenheit and meeting ASTM Specification D 1655 and Military Specifications MILT5624P and MILT83133D (Grades JP5 and JP8). It is used for commercial and military turbojet and turboprop aircraft engines.

Kyoto Protocol: The result of negotiations at the third Conference of the Parties (COP3) in Kyoto, Japan, in December of 1997. The Kyoto Protocol sets binding greenhouse gas emissions targets for countries that sign and ratify the agreement. The gases covered under the Protocol include carbon dioxide, methane, nitrous oxide, hydrofluorocarbons (HFCs), perfluorocarbons (PFCs) and sulfur hexafluoride.

Ketonealcohol (cyclohexanol): An oily, colorless, hygroscopic liquid with a camphorlike odor. Used in soapmaking, dry cleaning, plasticizers, insecticides, and germicides.

Leachate: The liquid that has percolated through the soil or other medium.

Lignite: The lowest rank of coal, often referred to as brown coal, used almost exclusively as fuel for steamelectric power generation. It is brownishblack and has a high inherent moisture content, sometimes as high as 45 percent The heat content of lignite ranges from 9 to 17 million Btu per ton on a moist, mineralmatterfree basis. The heat content of lignite consumed in the United States averages 13 million Btu per ton, on the asreceived basis (i.e., containing both inherent moisture and mineral matter).

Liquefied petroleum gases: A group of hydrocarbonbased gases derived from crude oil refining or natural gas fractionation. They include ethane, ethylene, propane, propylene, normal butane, butylene, isobutane, and isobutylene. For convenience of transportation, these gases are liquefied through pressurization.

Lubricants: Substances used to reduce friction between bearing surfaces, or incorporated into other materials used as processing aids in the manufacture of other products, or used as carriers of other materials. Petroleum lubricants may be produced either from distillates or residues. Lubricants include all grades of lubricating oils, from spindle oil to cylinder oil to those used in greases.

Methane: A colorless, flammable, odorless hydrocarbon gas (CH_4) which is the major component of natural gas. It is also an important source of hydrogen in various industrial processes. Methane is a greenhouse gas. See also *Greenhouse gases.*

Methanogens: Bacteria that synthesize methane, requiring completely anaerobic conditions for growth.

Methanol: A light alcohol that can be used for gasoline blending. See oxygenate.

Methanotrophs: Bacteria that use methane as food and oxidize it into carbon dioxide.

Methyl chloroform (trichloroethane): An industrial chemical (CH_3CCl_3) used as a solvent, aerosol propellant, and pesticide and for metal degreasing.

Methyl tertiary butyl ether (MTBE): A colorless, flammable, liquid oxygenated hydrocarbon containing 18.15 percent oxygen.

Methylene chloride: A colorless liquid, nonexplosive and practically nonflammable. Used as a refrigerant in centrifugal compressors, a solvent for organic materials, and a component in nonflammable paint removers.

Mole: The quantity of a compound or element that has a weight in grams numerically equal to its molecular weight. Also referred to as gram molecule or gram molecular weight.

Montreal Protocol: The Montreal Protocol on Substances that Deplete the Ozone Layer (1987). An international agreement, signed by most of the industrialized nations, to substantially reduce the use of chlorofluorocarbons (CFCs). Signed in January 1989, the original document called for a 50percent reduction in CFC use by 1992 relative to 1986 levels. The subsequent London Agreement called for a complete elimination of CFC use by 2000. The Copenhagen Agreement, which called for a complete phaseout by January 1, 1996, was implemented by the U.S. Environmental Protection Agency.

Motor gasoline (finished): A complex mixture of relatively volatile hydrocarbons with or without small quantities of additives, blended to form a fuel suitable for use in sparkignition engines. Motor gasoline, as defined in ASTM Specification D 4814 or Federal Specification VVG1690C, is characterized as having a boiling range of 122 to 158 degrees Fahrenheit at the 10 percent recovery point to 365 to 374 degrees Fahrenheit at the 90 percent recovery point. "Motor Gasoline" includes conventional gasoline; all types of oxygenated gasoline, including gasohol; and reformulated gasoline, but excludes aviation gasoline. Note: Volumetric data on blending components, such as oxygenates, are not counted in data on finished motor gasoline until the blending components are blended into the gasoline.

Multiple cropping: A system of growing several crops on the same field in one year.

Municipal solid waste: Residential solid waste and some nonhazardous commercial, institutional, and industrial wastes.

Naphtha less than 401 degrees Fahrenheit: A naphtha with a boiling range of less than 401 degrees Fahrenheit that is intended for use as a petrochemical feedstock.

Also see *Pet rochemical feed stocks.*

Naphthatype jet fuel: A fuel in the heavy naphtha boiling range having an average gravity of 52.8 degrees API, 20 to 90 percent distillation temperatures of 290 degrees to 470 degrees Fahrenheit, and meeting Military Specification MILT5624L (Grade JP4). It is used primarily for military turbojet and turboprop aircraft engines because it has a lower freeze point than other aviation fuels and meets engine requirements at high altitudes and speeds.

Natural gas: A mixture of hydrocarbons and small quantities of various nonhydrocarbons in the gaseous phase or in solution with crude oil in natural underground reservoirs.

Natural gas liquids (NGLs): Those hydrocarbons in natural gas that are separated as liquids from the gas. Includes natural gas plant liquids and lease condensate.

Natural gas, pipeline quality: A mixture of hydrocarbon compounds existing in the gaseous phase with sufficient energy content, generally above 900 Btu, and a small enough share of impurities for transport through commercial gas pipelines and sale to endusers.

Nitrogen oxides (NO_x): Compounds of nitrogen and oxygen produced by the burning of fossil fuels.

Nitrous oxide (N_2O): A colorless gas, naturally occurring in the atmosphere. Nitrous oxide has a 100year Global Warming Potential of 310.

Nonmethane volatile organic compounds (NMVOCs): Organic compounds, other than methane, that participate in atmospheric photochemical reactions.

Octane: A flammable liquid hydrocarbon found in petroleum. Used as a standard to measure the antiknock properties of motor fuel.

Oil reservoir: An underground pool of liquid consisting of hydrocarbons, sulfur, oxygen, and nitrogen trapped within a geological formation and protected from evaporation by the overlying mineral strata.

Organic content: The share of a substance that is of animal or plant origin.

Organic waste: Waste material of animal or plant origin.

Oxidize: To chemically transform a substance by combining it with oxygen.

Oxygenates: Substances which, when added to gasoline, increase the amount of oxygen in that gasoline blend. Ethanol, Methyl Tertiary Butyl Ether (MTBE), Ethyl Tertiary Butyl Ether (ETBE), and methanol are common oxygenates.

Ozone: A molecule made up of three atoms of oxygen. Occurs naturally in the stratosphere and provides a protective layer shielding the Earth from harmful ultraviolet radiation. In the troposphere, it is a chemical oxidant, a greenhouse gas, and a major component of photochemical smog.

Ozone precursors: Chemical compounds, such as carbon monoxide, methane, nonmethane hydrocarbons, and nitrogen oxides, which in the presence of solar radiation react with other chemical compounds to form ozone.

Paraffinic hydrocarbons: Straightchain hydrocarbon compounds with the general formula C_nH_{2n+2}.

Perfluorocarbons (PFCs): A group of manmade chemicals composed of one or two carbon atoms and four to six fluorine atoms, containing no chlorine. PFCs have no commercial uses and are emitted as a byproduct of aluminum smelting and semiconductor manufacturing. PFCs have very high 100year Global Warming Potentials and are very longlived in the atmosphere.

Perfluoromethane: A compound (CF_4) emitted as a byproduct of aluminum smelting.

Petrochemical feedstocks: Chemical feedstocks derived from petroleum principally for the manufacture of chemicals, synthetic rubber, and a variety of plastics.

Petroleum: A broadly defined class of liquid hydrocarbon mixtures. Included are crude oil, lease condensate, unfinished oils, refined products obtained from the processing of crude oil, and natural gas plant liquids. Note: Volumes of finished petroleum products include nonhydrocarbon compounds, such as additives and detergents, after they have been blended into the products.

Petroleum coke: See *Coke (petroleum).*

Photosynthesis: The manufacture by plants of carbohydrates and oxygen from carbon dioxide and water in the presence of chlorophyll, with sunlight as the energy source. Carbon is sequestered and oxygen and water vapor are released in the process.

Pig iron: Crude, highcarbon iron produced by reduction of iron ore in a blast furnace.

Pipeline, distribution: A pipeline that conveys gas from a transmission pipeline to its ultimate consumer.

Pipeline, gathering: A pipeline that conveys gas from a production well/field to a gas processing plant or transmission pipeline for eventual delivery to enduse consumers.

Pipeline, transmission: A pipeline that conveys gas from a region where it is produced to a region where it is to be distributed.

Planetary albedo: The fraction of incident solar radiation that is reflected by the Earthatmosphere system and returned to space, mostly by backscatter from clouds in the atmosphere.

Pneumatic device: A device moved or worked by air pressure.

Polystyrene: A polymer of styrene that is a rigid, transparent thermoplastic with good physical and electrical insulating properties, used in molded products, foams, and sheet materials.

Polyvinyl chloride (PVC): A polymer of vinyl chloride. Tasteless. odorless, insoluble in most organic solvents. A member of the family vinyl resin, used in soft flexible films for food packaging and in molded rigid products, such as pipes, fibers, upholstery, and bristles.

Postmining emissions: Emissions of methane from coal occurring after the coal has been mined, during transport or pulverization.

Radiative forcing: A change in average net radiation at the top of the troposphere (known as the tropopause) because of a change in either incoming solar or exiting infrared radiation. A positive radiative forcing tends on average to warm the earth's surface; a negative radiative forcing on average tends to cool the earth's surface. Greenhouse gases, when emitted into the atmosphere, trap infrared energy radiated from the earth's surface and therefore tend to produce positive radiative forcing. See *Greenhouse gases.*

Radiatively active gases: Gases that absorb incoming solar radiation or outgoing infrared radiation, affecting the vertical temperature profile of the atmosphere. See *Radiative forcing.*

Ratoon crop: A crop cultivated from the shoots of a perennial plant.

Redox potential: A measurement of the state of oxidation of a system.

Reflectivity: The ratio of the energy carried by a wave after reflection from a surface to its energy before reflection.

Reforestation: Replanting of forests on lands that have recently been harvested or otherwise cleared of trees.

Reformulated gasoline: Finished motor gasoline formulated for use in motor vehicles, the composition and properties of which meet the requirements of the ref ormulated gasoline regulations promulgated by the U.S. Environmental Protection Agency under Section 211(k) of the Clean Air Act. Note: This category includes oxygenated fuels program reformulated gasoline (OPRG) but excludes reformulated gasoline blendstock for oxygenate blending (RBOB).

Renewable energy re Sources**:** Energy resources that are naturally replenishing but flowlimited. They are virtually inexhaustible in duration but limited in the amount of energy that is available per unit of time. Renewable energy resources include: biomass, hydro, geothermal, solar, wind, ocean thermal, wave action, and tidal action.

Residual fuel oil: A general classification for the heavier oils, known as No.5 and No.6 fuel oils, that remain after the distillate fuel oils and lighter hydrocarbons are distilled away in refinery operations. It conforms to ASTM Specifications D 396 and D 975 and Federal Specification VVF815C. No. 5, a residual fuel oil of medium viscosity, is also known as Navy Special and is defined in Military Specification MILF859E, including Amendment 2 (NATO Symbol F770). It is used in steampowered vessels in government service and inshore powerplants. No. 6 fuel oil includes Bunker C fuel oil and is used for the production of electric power, space heating, vessel bunkering, and various industrial purposes.

Rumen: The large first compartment of the stomach of certain animals in which cellulose is broken down by the action of bacteria.

Sample: A set of measurements or outcomes selected from a given population.

Sequestration: See *Carbon sequestration.*

Septic tank: A tank in which the solid matter of continuously flowing sewage is disintegrated by bacteria.

Sinter: A chemical sedimentary rock deposited by precipitation from mineral waters, especially siliceous sinter and calcareous sinter.

Sodium silicate: A greywhite powder soluble in alkali and water, insoluble in alcohol and acid. Used to fireproof textiles, in petroleum refining and corrugated paperboard manufacture, and as an egg preservative. Also referred to as liquid gas, silicate of soda, sodium metasilicate, soluble glass, and water glass.

Sodium tripolyphosphate: A white powder used for water softening and as a food additive and texturizer.

Stabilization lagoon: A shallow artificial pond used for the treatment of wastewater. Treatment includes removal of solid material through sedimentation, the decomposition of organic material by bacteria, and the removal of nutrients by algae.

Still gas (refinery gas): Any form or mixture of gases produced in refineries by distillation, cracking, reforming, and other processes. The principal constituents are methane, ethane, ethylene, normal butane, butylene, propane, propylene, etc. Still gas is used as a refinery fuel and a petrochemical feedstock. The conversion factor is 6 million Btu per fuel oil equivalent barrel.

Stratosphere: The region of the upper atmosphere extending from the tropopause (8 to 15 kilometers altitude) to about 50 kilometers. Its thermal structure, which is determined by its radiation balance, is generally very stable with low humidity.

Stripper well: An oil or gas well that produces at relatively low rates. For oil, stripper production is usually defined as production rates of between 5 and 15 barrels of oil per day. Stripper gas production would generally be anything less than 60 thousand cubic feet per day.

Styrene: A colorless, toxic liquid with a strong aromatic aroma. Insoluble in water, soluble in alcohol and ether; polymerizes rapidly; can become explosive. Used to make polymers and copolymers, polystyrene plastics, and rubber.

Subbituminous coal: A coal whose properties range from those of lignite to those of bituminous coal and used primarily as fuel for steamelectric power generation. It may be dull, dark brown to black, soft and crumbly, at the lower end of the range, to bright, jet black, hard, and relatively strong, at the upper end. Subbituminous coal contains 20 to 30 percent inherent moisture by weight. The heat content of subbituminous coal ranges from 17 to 24 million Btu per ton on a moist, mineralmatterfree basis. The heat content of subbituminous coal consumed in the United States averages 17 to 18 million Btu per ton, on the asreceived basis (i.e., containing both inherent moisture and mineral matter).

Sulfur dioxide (SO_2): A toxic, irritating, colorless gas soluble in water, alcohol, and ether. Used as a chemical intermediate, in paper pulping and ore refining, and as a solvent.

Sulfur hexafluoride (SF_6): A colorless gas soluble in alcohol and ether, and slightly less soluble in water. It is used as a dielectric in electronics. It possesses the highest 100year Global Warming Potential of any gas (23,900).

Sulfur oxides (SO_x): Compounds containing sulfur and oxygen, such as sulfur dioxide (SO_2) and sulfur trioxide (SO_3).

Tertiary amyl methyl ether ($(CH_3)_2(C_2H_5)COCH_3$): An oxygenate blend stock formed by the catalytic etherification of isoamylene with methanol.

Troposphere: The inner layer of the atmosphere below about 15 kilometers, within which there is normally a steady decrease of temperature with increasing altitude. Nearly all clouds form and weather conditions manifest themselves within this region. Its thermal structure is caused primarily by the heating of the earth's surface by solar radiation, followed by heat transfer through turbulent mixing and convection.

Uncertainty: A measure used to quantify the plausible maximum and minimum values for emissions from any source, given the biases inherent in the methods used to calculate a point estimate and known sources of error.

Vapor displacement: The release of vapors that had previously occupied space above liquid fuels stored in tanks. These releases occur when tanks are emptied and filled.

Ventilation system: A method for reducing methane concentrations in coal mines to nonexplosive levels by blowing air across the mine face and using large exhaust fans to remove methane while mining operations proceed.

Vessel bunkering: Includes sales for the fueling of commercial or private boats, such as pleasure craft, fishing boats, tugboats, and oceangoing vessels, including vessels operated by oil companies. Excluded are volumes sold to the U.S. Armed Forces.

Volatile organic compounds (VOCs): Organic compounds that participate in atmospheric photochemical reactions.

Volatile solids: A solid material that is readily decomposable at relatively low temperatures.

Waste flow: Quantity of a waste stream generated by an activity.

Wastewater: Water that has been used and contains dissolved or suspended waste materials.

Wastewater, domestic and commercial: Wastewater (sewage) produced by domestic and commercial establishments.

Wastewater, industrial: Wastewater produced by industrial processes.

Water vapor: Water in a vaporous form, especially when below boiling temperature and diffused (e.g., in the atmosphere).

Wax: A solid or semisolid material derived from petroleum distillates or residues by such treatments as chilling, precipitating with a solvent, or deoiling. It is a lightcolored, moreorless translucent crystalline mass, slightly greasy to the touch, consisting of a mixture of solid hydrocarbons in which the paraffin series predominates. Includes all marketable wax, whether crude scale or fully refined. The three grades included are microcrystalline, crystallinefully refined, and crystallineother. The conversion factor is 280 pounds per 42 U.S. gallons per barrel.

Weanling system: A cattle management system that places calves on feed starting at 165 days of age and continues until the animals have reached slaughter weight.

Wellhead: The point at which the crude (and/or natural gas) exits the ground. Following historical precedent, the volume and price for crude oil production are labeled as "wellhead," even though the cost and volume are now generally measured at the lease boundary. In the context of domestic crude price data, the term "wellhead" is the generic term used to reference the production site or lease property.

Wetlands: Areas regularly saturated by surface or groundwater and subsequently characterized by a prevalence of vegetation adapted for life in saturatedsoil conditions.

Wood energy: Wood and wood products used as fuel, including roundwood (cordwood), limbwood, wood chips, bark, sawdust, forest residues, charcoal, pulp waste, and spent pulping liquor..

Yearling system: A cattle management system that includes a stocker period from 165 days of age to 425 days of age followed by a 140day feedlot period.

In: Greenhouse Gases and Their Impact
Editor: Bernard F. Shulk, pp. 143-156

ISBN: 978-1-60021-392-2

Chapter 2

GLOBAL CLIMATE CHANGE: U.S. GREENHOUSE GAS EMISSIONS—STATUS, TRENDS AND PROJECTIONS*

John Blodgett and Larry Parker

SUMMARY

This report reviews U.S. emissions of greenhouse gases in the contexts both of domestic policy and of international obligations and proposals. On October 15, 1992, the United States ratified the United Nations Framework Convention on Climate Change (UNFCCC), which entered into force on March 21, 1994. This committed the United States to "national policies" to limit "its anthropogenic emissions of greenhouse gases," with a voluntary goal of returning "emissions of carbon dioxide [CO_2] and other greenhouse gases [methane (CH_4), nitrous oxide (N_2O),hydrofluorocarbons (HFCs), perfluorocarbons (PFCs), and sulfurhexafluoride (SF_6)]" at the "end of the decade" to "their 1990 levels."

Subsequently, in the 1997 Kyoto Protocol to the UNFCCC, the United States participated in negotiations that ended with agreement on further reductions that could become legally binding. The United States signed the Kyoto Protocol in 1998, but President Clinton did not send it to the Senate for advice and consent. President Bush has said that he rejects the Protocol, and former U.S. Environmental Protection Agency Administrator Christine Todd Whitman told reporters that the Administration would not be pursuing the UNFCCC commitment either. Instead, President Bush has proposed to shift the nation's climate change program from a goal of reducing emissions per se to a goal of reducing energy intensity — the amount of greenhouse gases emitted per unit of economic productivity. Under the proposal, the intensity, which has been declining for a number of years, would decline 18% between 2002 and 2012, as opposed to a 14% projected "business as usual" decline.

Meanwhile, the UNFCCC "end of the decade" deadline has passed and U.S. greenhouse gas emissions continue on an upward trend, though with dips in 1991and in 2001, attributed mostly to economic slowdowns. Based on historical data, 2001 emissions were about 13% in excess of the UNFCCC goal. Overall, from 1990 to 2001, U.S. greenhouse gas emissions

* Excerpted from CRS Report 98-235 ENR . Updated August 15, 2003

(weighted by global warming potential) have increased an average of about 1.1% per year. Projections suggest that U.S. emissions will continue to rise for at least the next decade. Reversing the upward trend in greenhouse gas emissions would represent an extraordinary technical and political challenge to U.S. energy and environmental policy.
This report will be updated as necessary.

GLOBAL CLIMATE CHANGE: U.S. GREENHOUSE GAS EMISSIONS— STATUS, TRENDS, AND PROJECTIONS

This report reviews U.S. emissions of greenhouse gases in the contexts both of domestic policy and of international obligations and proposals. On October 15, 1992, the United States ratified the United Nations Framework Convention on Climate Change (UNFCCC), which entered into force on March 21, 1994. By this action, the nation made a legally non-binding commitment to "national policies" to limit "its anthropogenic emissions of greenhouse gases," which are believed to pose a risk of global climate change.[1] The goal was to return "these anthropogenic emissions of carbon dioxide and other greenhouse gases" at the "end of the decade" to "their 1990 levels." This goal was voluntary, to "demonstrate that developed countries are taking the lead in modifying longer-term trends in anthropogenic emissions consistent with the objective of the Convention."[2] The Convention established standards for inventorying and reporting greenhouse gas emissions.

Subsequently, the United States participated in negotiating the Kyoto Protocol to the UNFCCC.[3] Under the Kyoto Protocol, the United States would have made a legally binding agreement that for the 5-year period 2008-2012, it would reduce its average annual aggregate carbon-equivalent emissions of six gases by 7% below specified baseline years.[4] However, while President Clinton signed the Protocol in 1998, he did not send it to the Senate for its advice and consent; and President George W. Bush has said that he has no intention of pursuing the Kyoto Protocol — that it is "fatally flawed."[5]

Nonetheless, other nations continue efforts to implement the Kyoto Protocol, and the United States remains obligated under the UNFCCC to inventory its emissions of greenhouse gases and to pursue voluntary reductions. However, as described below, President Bush has announced a voluntary program to reduce the intensity of U.S. greenhouse gas emissions per unit of economic productivity.[6] Based on this initiative, according to former Environmental Protection Agency Administrator Christine Todd Whitman, "the Bush Administration does not intend to pursue emissions cuts it committed to" in the UNFCCC.[7]

It is in this context, then, that this report sets out the baseline emissions that the United States has established and portrays trends in emissions over the past decade;[8] notes the current status of U.S. emissions as compared to the UNFCCC goals; and reviews projections and, as a point of reference, compares them to the Kyoto Protocol emissions commitments. (Assessing that prospective commitment does not imply that the United States should reduce emissions, only what would be required *if* it were to join that agreement to reduce emissions.)

Table 1. U.S. Greenhouse Gas Emissions (MMTCE), 1990-2001

Gas	1990	1991	1992	1993	1994	1995	1996	1997	1998	1999	2000	2001
CO2	1,365	1,354	1,381	1,412	1,438	1,455	1,504	1,526	1,531	1,549	1,604	1,580
CO2 (sinks)[a]	(293)	(289)	(291)	(291)	(292)	(290)	(289)	(229)	(226)	(229)	(228)	(229)
CH4	176	176	176	174	176	177	174	172	170	168	167	165
N2O	108	110	113	113	120	118	120	120	119	117	117	116
HFCs, PFCs, SF6	26	24	25	26	26	27	31	32	35	33	33	30
Total emissions	1,675	1,664	1,694	1,726	1,760	1,777	1,829	1,850	1,855	1,867	1,921	1,892
Net emissions	1,382	1,375	1,403	1,435	1,468	1,487	1,540	1,621	1,629	1,638	1,693	1,663

Source: Environmental Protection Agency, *Inventory of U.S. Greenhouse Gas Emissions and Sinks: 1990-2001* (April 2003), EPA 430R03004. (Data for 1991 1994 provided by EPA.) Following international convention, EPA presents the data in teragrams of CO_2 equivalent; CRS has converted the figures to million metric tons of carbon equivalent (MMTCE), a metric that is more familiar to most U.S. policymakers.

[a] Land-use changes and forestry sinks that sequester carbon; included in net emissions total only.

In what follows, figures for emissions are point estimates[9] and rounded to millions of metric tons of carbon equivalents (MMTCE[10]). As will be discussed, the data are of varying robustness and may be subject to adjustments. The data for CO_2, which accounts for over 80% of domestic greenhouse gas emissions, are the most robust, being largely based on comprehensive fuel use data. Subsumed estimates and uncertainties in projected emissions have greater effect the further into the future one looks. But even allowing for these imprecisions, the *trendline* of aggregate U.S. emissions of greenhouse gases is clear: over the decade of the 1990s emissions trended upward, and they are projected to continue to rise in the future.

U.S. Greenhouse Gas Emissions and Baselines

Pursuant to the United Nations Framework Conventionon Climate Change, the United States has published "national inventories of anthropogenic emissions by sources and removals by sinks of all greenhouse gases not controlled by the Montreal Protocol, using comparable methodologies ... agreed upon by the Conference of the Parties."[11] (See **Table 1**.)

The Environmental Protection Agency (EPA) publishes the official emissions data annually.[12] The United States also from time to time reports on emissions and explains its climate change programs in the *Climate Action Report (CAR)* to the United Nations; the third *CAR* was published in 2002.[13]

The U.S. baselines for the UNFCCC and the Kyoto Protocol are shown in **Table 2**. For the UNFCCC commitment, the baseline is 1990 emissions, or 1,675 MMTCE; if the United States had acceded to the Kyoto targets, the baseline would have been 1,676 MMTCE, a negligible difference. By definition, sinks are *not* included in calculating the baselines.

Table 2. U.S. Baseline Year Greenhouse Gas Emissions

Greenhouse Gas	Baseline Year		Emissions (MMTCE)	
Carbon dioxide (CO_2)	1990		1,365	
Methane (CH_4)	1990		176	
Nitrous oxide (N_2O)	1990		108	
Hydrofluorocarbons (HFCs)	1990 UNFCCC	1995 Kyoto	26 UNFCCC	27 Kyoto
Perfluorocarbons (PFCs)				
Sulfur hexafluoride (SF_6)				
Total			1,675	1,676

Source: EPA, Inventory of U.S. Greenhouse Gas Emissions and Sinks: 1990 -2001) April 2003, EPA 430R03004.

The emissions baselines shown in **Table 2** are not immutable. Each annual report includes updated estimates based on methodological and data revisions, although such changes are usually small. Revisions are discussed at some length in each report. The criteria for calculating emissions agreed upon by the Conference of Parties hinge on both current technical knowledge and policy judgments. New technical information can change factors, for example concerning calculation of greenhouse gas equivalents; and policy judgments can be adjusted, for example concerning the time frame for calculating effects. In addition, a few technical issues remain unresolved, for example in assigning emissions from fuels burned in international travel. However, any changes tend to be modest, seldom affecting totals much more than plusorminus1%, except for sequestration figures, which have been subject to larger changes.[14]

Table 3. U.S. Greenhouse Gas Intensity (1990-2001)

	GDP (billions of chained (1996) dollars)	Greenhouse Gas Intensity (metric tons carbon equivalent per million $ GDP)	
		Total Emissions	Net Emissions
1990	6,708	250	206
1991	6,677	249	206
1992	6,880	246	204
1993	7,063	244	203
1994	7,348	240	200
1995	7,544	236	197
1996	7,813	234	197
1997	8,160	227	199
1998	8,509	218	184
1999	8,859	211	185
2000	9,191	209	184
2001	9,214	205	180

Source: Table 1; *Economic Report of the President*, February 2003, Table B-2; CRS calculations.

Besides actual quantities of emissions, an alternative measure of a nation's contribution to global warming is "greenhouse gas intensity of the economy" — that is, emissions per unit of gross domestic product (GDP). In effect, this measure focuses on the efficiency of the economy in terms of greenhouse gas emissions: the more efficient, the fewer emissions per dollar of economic output and thus the lower the "greenhouse gas intensity." For the United States, greenhouse gas intensity has been declining since at least the 1980s; for the 1990s, the decline in intensity was about 10%, based on net emissions[15] (see **Table 3**).

Emissions Projections

Projecting greenhouse gas emissions involves modeling the nation's economic growth and activity, with special attention to variables affecting fossil fuel combustion. The modeling also depends on assumptions about energy policy directions. If reducing emissions becomes a goal, then projections become subject to the outcome of unresolved issues in how the emissions reductions goals might be met.

For example, the major source of CO_2 emissions, fossil fuel combustion, is influenced by overall economic activity and growth as well as by energy policy decisions such as development of non-carbon based substitutes, the rate of adoption of energy efficient technologies, and the retirement rate of nuclear facilities, among others. These policy factors are difficult to predict in the absence of a concrete climate change policy.[16] The climate change plan proposed by President Bush in February 2002 provides some new policy directions, but many elements depend on congressional action (e.g., for funding) or voluntary private sector initiatives, making projections of their impact problematic.

The Climate Action Report Projection

The third U.S. *Climate Action Report* (*CAR 2002*) projects greenhouse gas emissions at 5-year intervals through 2020. For this report, the projections are followed only to 2010 (see **Figures 1 and 2**), because of the difficulties in projecting into the more distant future. Also, 2010 provides a basis for evaluating a relationship to the Kyoto Protocol targets.

The *CAR 2002* estimate for aggregate gross greenhouse emissions[17] in 2010 is 2,213MMTCE(see **Figure3**). The President's 2002 initiative to reduce greenhouse gas intensity proposes a series of policy initiatives that it estimates "will achieve 100 million tons of reduced emissions in 2012."[18] Extrapolating between the *CAR 2002* projections from 2010 to 2015 (a 1.9% annual growth rate), the 2012 projected emission level would be 2,298 MMTCE. The President's initiative suggesting a decline in emissions of 100 million tons in 2012 would reduce this to 2,198 MMTCE, representing a reduction of about 4.4% from "business as usual" gross greenhouse emissions in that year. In addition, largely separate from federal activities, a number of state and local governmental initiatives, as well as a variety of private sector activities, are underway to address greenhouse gas emissions.

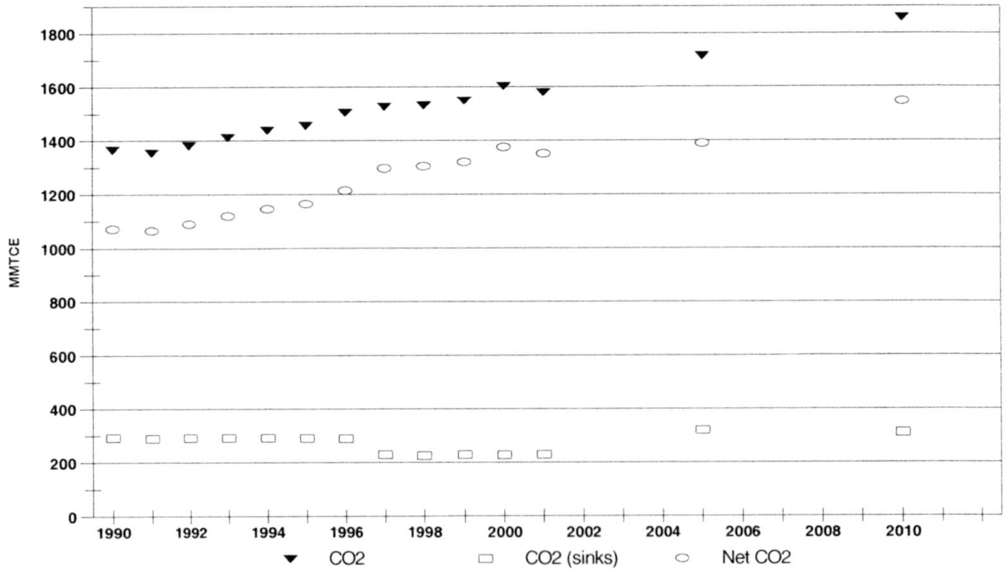

Sources: Historical data (through 2001): EPA, *Inventory of U.S. Greenhouse Gas Emissions and Sinks: 1990 -2001*, April 2003, EPA 430R03004, pp. ES-2 -ES-4. Projections (to 2010): U.S. Department of State, *Climate Action Report 2002*, May 2002. [Data converted to MMTCE by CRS.]

Figure 1. U.S. Emissions of CO_2: Historical and Projected (2005, 2010)

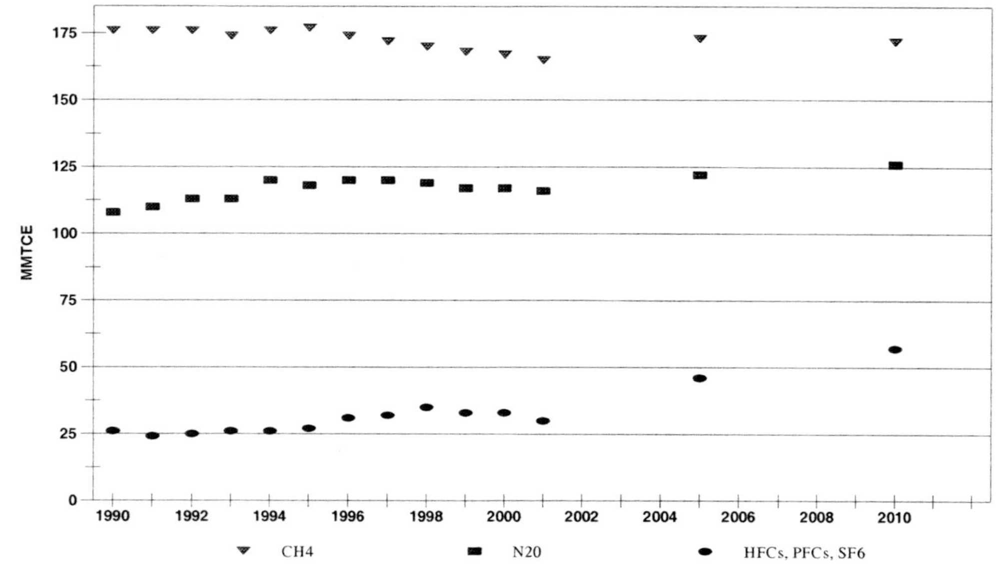

Sources: Historical data (through 2001): EPA, *Inventory of U.S. Greenhouse Gas Emissions and Sinks: 1990 -2001*, April 2003, EPA 430R03004, pp. ES-2 -ES-4. Projections (to 2010): U.S. Department of State, *Climate Action Report 2002*, May 2002. [Data converted to MMTCE by CRS.]

Figure 2. U.S. Emissions of CH_4,N_20, and HFCs, PFCs, and SF_6: Historical and Projected (2005, 2010)

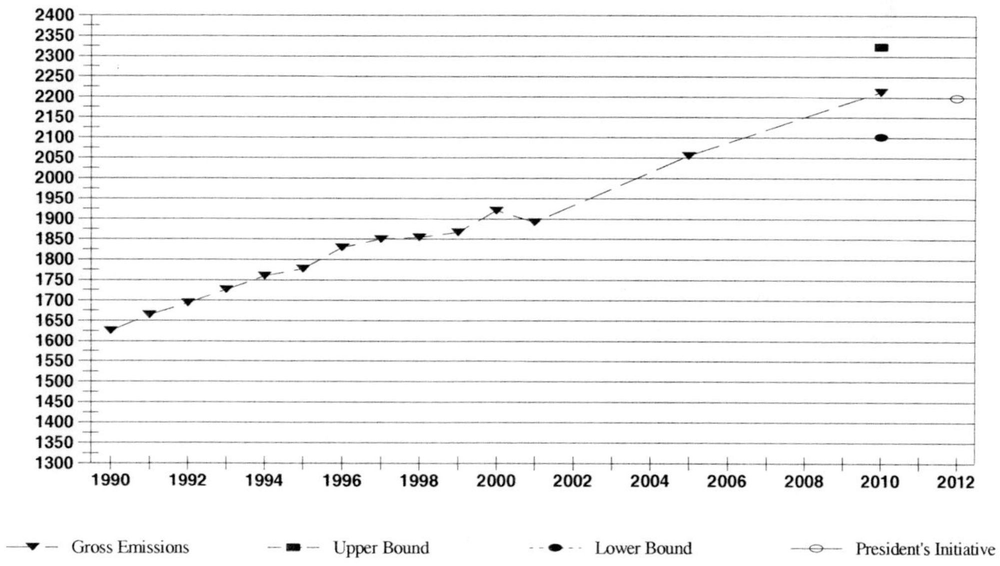

Sources: Historical data (through 2001): EPA, *Inventory of U.S. Greenhouse Gas Emissions and Sinks: 1990 -2001*, April 2003, EPA 430R03004, pp. ES-2 -ES-4. Projections (to 2010): U.S. Department of State, *Climate Action Report 2002*, May 2002. [Data converted to MMTCE by CRS.] Upper and lower bound equal + 5% and -5%, as discussed in text. President's initiative is from documents on the Administration plan at [http://www.whitehouse.gov/news/releases/2002/02/climatechange.html].

Figure 3. U.S. Aggregate Gross Emissions of Six Greenhouse Gases:Historical and Projected (2005, 2010) (MMTCE)

Effect on Projections of Varying Assumptions

CAR 2002 only makes point estimates, but some sense of the implications of varying assumptions that affect the estimates can be gleaned from examining an alternative source of CO_2 emissions data, the Energy Information Agency's (EIA's) *Annual Energy Outlook* series.[19] (Because of minor differences in data calculation and presentation, EIA's annual emissions figures differ slightly from EPA's.)

The EIA report's projections of CO_2 emissions include sensitivity analyses to various changes in assumptions, and since CO_2 from fuel combustion accounts for about 80% of U.S. greenhouse gas emissions, the analysis is a reasonable test of the projections. The assumptions EIA examines include economic growth, technological innovation, oil prices, electricity demand, and others. The first two, economic growth and technological innovation, have the greatest effect on variance in projections of CO_2emissions(see **Table4**). For 2010, compared to EIA's "reference case" (which is equivalent to a "business as usual" case), low economic growth wouldreduceprojectedemissionsbyabout2%,[20] while high economic growth would increase projected emissions by about 3%. Compared to the reference case that assumes anticipated technological developments, static technology would result in emissions rising about 2%, while a "high-technology" case is projected to reduce emissions about 2%. The point reference case — *CAR's* point estimate — effectively assumes these veral variances affecting emissions cancel out. But if all the variances increasing emissions prove true and cumulative, then projected emissions for 2010 could be 5% or more higher than the point estimate; conversely, if all the variances decreasing emissions prove true and cumulative, emissions could be 5% or more lower.

Table 4. Impact of Economic Assumptions on Projections of CO_2 Emissions

Case Comparisons	Change in CO2 Emissions from Fuel Use, 2010		
	Low economic growth MMTCE (%)	Reference case MMTCE	High economic growth MMTCE (%)
Economic Growth	1,759 (-2%)	1,800	1,852 (+3%)
	2003 Technology	Reference Case	High Technology
Integrated Technology	1,831 (+2%)	1,800	1,759 (-2%)

Source: EIA, *Annual Energy Outlook 2003* (January 2003) DOE/EIA-0383(2003), pp. 174, 218.

Some studies suggest that even greater variance in projections is possible—for example, that new energy-efficient technologies could be deployed more quickly than generally assumed if appropriate policies were instituted. A November 2000 DOE study, commonly called the "New 5-Lab Study," shows that energy efficiency gains in the transportation, industry, commercial, and residential sectors could reduce emissions from the "business as usual" scenario.[21] The "business as usual" scenario in this study is very similar to EIA's reference case, though it projects somewhat smaller emissions in 2010 (1,769 MMTCE from fossil fuel combustion, compared to EIA's most recent projection of 1,800). The study compares "moderate" and "advanced" scenarios "that are defined by policies that are consistent with increasing levels of public commitment and political resolve to solving the nation's energy-related challenges." Policies examined include "fiscal incentives, voluntary programs, regulations, and research and development."[22]

Under the "moderate scenario," energy efficiency is improved through such policies as expanded labeling, new efficiency standards, tax credits, and cost-shared R&D; renewable energy grows more rapidly than in the "business as usual" scenario, and a higher proportion of nuclear power is retained. Under the "advanced scenario," which has more aggressive demand-and supply-side policies and a doubling of R&D, a federally sponsored carbon trading system is announced in 2002 and implemented in 2005 with a clearing equilibrium price of $50 per ton of carbon.[23] The results of this analysis are shown in **Table 5**.

Table 5. Impact of Technology/Efficiency Assumptions onProjections of CO_2 Emissions

Case Comparisons	Total CO2 Emissions from Fuel Use, 2010 (MMTCE)
"Business as Usual" (BAU)	1,769
Moderate Scenario	1,684 (-5% from BAU)
Advanced Scenario	1,467 (-17% from BAU)

Source: DOE, Interlaboratory Working Group, *Scenarios for a Clean Energy Future*, ORNL/CON-476andLBNL-44029(OakRidge,TN:OakRidgeNationalLaboratory;Berkeley,CA: Lawrence Berkeley National Laboratory, 2000), Table 1.8, p. 1.18.

This "New 5-Lab Study" thus suggests that if specified policies were adopted, emissions could be considerably lower than even EIA's high-technology scenario indicates, by as much as 17% compared to EIA's high-technology reduction in emissions of about 2%. EPA and the Department of Energy (DOE) have underway a number of programs to foster the development and deployment of energy-efficient technologies.[24]

The President's greenhouse gas intensity reduction initiative is a new variable affecting projections. It would have the effect of reducing anticipated emissions below "business as usual" levels in the future (see **Figure 3**); however, the initiative does not reflect the level of aggressiveness assumed by the "New 5-Lab Study" for policy interventions to achieve its "advanced scenario" for rapid penetration of energy-efficient technologies.

Status of Emissions Relative to Goals

Under the UNFCCC, the United States committed to the voluntary goal of holding greenhouse gas emissions at the end of the 1990s to their 1990 levels. If the United States had met this goal, its greenhouse gas emissions for 2000 would have been 1,675 MMTCE. However, U.S. emissions in 2000 were 1,921 MMTCE (not counting sinks). These figures indicate that in 2000, the nation was exceeding its UNFCCC greenhouse gas emissions commitment by 246 MMTCE, or nearly 15%.

If the United States had acceded to the Kyoto Protocol, its greenhouse gases emissions target for the period 2008-2012 would have been 5 times 93% of the 1,676 MMTCE baseline, or 1,559 MMTCE on average per year for the period. This hypothetical goal would imply reductions equal to the difference between the goal and what would be "business as usual emissions" for the period 2008-2012. Based on the *CAR* projection that emissions will be 2,213 MMTCE in 2010, the average annual reduction that would be necessary for the United States to meet the Kyoto target of 1,559 MMTCE per year for 2008-2012 would be 654 MMTCE per year, or about 30% below the estimated level of "business as usual" emissions. Higher than base case economic growth or lower penetration of energy-efficient technologies would mean that emissions would be even higher (and reductions necessary to meet a goal like Kyoto greater). Slower economic growth, or faster penetration of energy-efficient technologies as suggested by the 5-Lab Study, would decrease emissions (and hence reductions to meet a goal).

The President's greenhouse gas initiative has the goal of reducing, through voluntary activities, the intensity of *net* greenhouse gas emissions per unit of economic activity by 18% over the next 10 years; this compares to a projected "business as usual" decline in intensity of 14% for the period — compared to a decline during the 1990s of about 10% (see **Table 3**). According to the White House announcement, this goal means that the current (2002) intensity of 183 metric tons of carbon emissions per million dollars of GDP would fall to 151 MMTCE per million dollars of GDP in 2012 (see **Figure 4**). [25] At the anticipated increased rate of intensity decline, total emissions would decline 100 MMTCE below "business as usual" emissions (although the absolute amount of emissions would continue to rise). It is too early to assess progress toward the Administration's goal of diminishing greenhouse gas intensity.

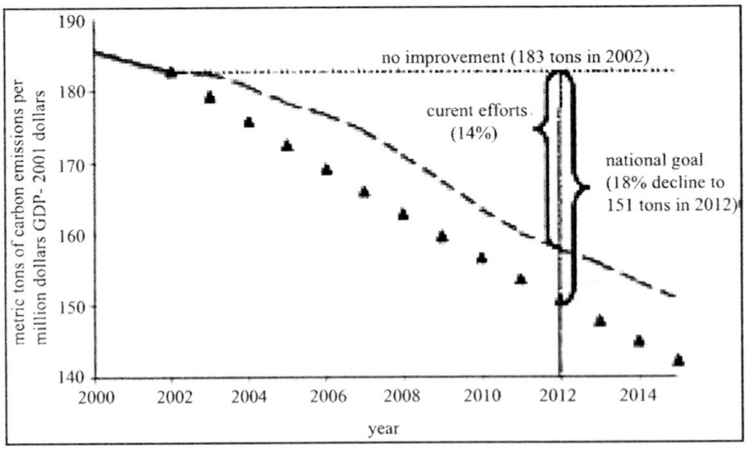

Source: White House, *Global Climate Change Policy Book*, Feb. 14, 2002, at [http://www.epa.gov/globalwarming/publications/actions/us_position/bush_gccp_021402.pdf].

Figure 4. Greenhouse Gas Intensity:President's Initiative

These projected emissions levels (and any implied reductions) are gross estimates and do not take sinks into account (except for the intensity projection). As previously noted, the baseline could be revised, at least slightly. More important, such projections depend on assumptions about economic trends as well as about policy actions at the local, domestic, and international levels. However, whatever the assumptions, the trend in total emissions experienced over the past decade and projected for the next decade is clearly upward, while the UNFCCC goal was for stabilization and the Kyoto Protocol calls for emissions levels of developed nations to decline.

Additional Considerations in Assessing Possible Reductions

If one is concerned about assessing the implications of possible reduction requirements in the future, two further factors must be considered. One is sequestration, which removes CO_2 from the atmosphere, thereby reducing gross emissions. The second is a series of proposed trading mechanisms that could allow a country to take credit for reductions it sponsors in other countries. The United States was a strong supporter of including both these variables in the Kyoto Protocol. Sequestration could directly diminish a country's reduction requirement; trading does not change a reduction requirement, but it could affect costs and who would actually achieve the reductions.

Carbon Sequestration. Atmospheric greenhouse gas levels are affected not only by emissions, but also by carbon sinks — processes that remove and sequester carbon from the atmosphere. Activities that affect sequestration include farming and forestry practices. For example, a positive net growth of trees removes carbon from the atmosphere; clearing forests typically releases carbon. **Table 1**,"U.S.

Greenhouse Gas Emissions, 1990 -2001," includes figures for carbon sequestration from land-use activities and forestry, which are the difference between "total emissions" and "net emissions."

The UNFCCC states that signatory nations shall commit to "promote sustainable management, and promote and cooperate in the conservation and enhancement, as appropriate, of sinks and reservoirs of all greenhouse gases ... , including biomass, forests and oceans as well as other terrestrial, coastal and marine ecosystems" (Article 4(1)(d)).

The Kyoto Protocol also would provide that sinks can be taken into account in calculating a nation's emissions and its reduction obligation. "The net changes in greenhouse gas emissions from sources and removals by sinks resulting from direct human-induced land-use change and forestry activities, limited to afforestation, reforestation, and deforestation since 1990, measured as verifiable changes in stocks ... shall be used to meet" the 2008-2012 commitments (Article 3(3)). In general, then, a net increase in human-induced carbon sequestration from forestry practices between 1990 and 2008-2012 would be subtracted from emissions during the period, thereby reducing the amount of actual emissions that will have to be curtailed. Conversely, net negative sequestration from forestry practices would be added to the emissions that will have to be reduced.

Just how this calculation would be done is not prescribed in the Protocol, and disagreements on how much carbon sequestration could be counted toward a nation's reduction obligations were debated through several subsequent conferences. In July 2001, the Sixth Conference of Parties in Bonn (COP6) agreed to limits on sequestration activities that could be credited against the Protocol's reduction requirements. Although the United States chose not to participate in these proceedings, the Conference stated in a footnote[26] that under the methodology agreed upon, the United States could take credit for net increases of sequestration of up to 28 million metric tons per year.

Emissions Trading. Emissions trading, strongly supported by the United States in the Kyoto negotiations, derives from the principle of economic efficiency — that reductions, if necessary, should be achieved at the lowest cost. Trading mechanisms thus are designed to allow low-cost reductions to substitute for higher-cost ones. The idea is that a country could achieve its reduction goal not only by reducing its domestic emissions, but also by reducing emissions elsewhere. Trading does not actually reduce a nation's reduction requirement, but it does allow it to contract for and to count reductions elsewhere that are cheaper to achieve than domestic ones.

The Kyoto Protocol provides for emissions trading mechanisms[27] that can be used to "supplement" domestic reductions; this offers the possibility that actual domestic greenhouse gas reductions achieved by a party to the Kyoto Protocol will be less than the party's actual commitment. Some portion of the reduction requirement could be shifted elsewhere. The Clinton Administration argued that emission trading would be critical to U.S. compliance with Kyoto;[28] a Clinton Administration economic analysis suggested that U.S. compliance costs would drop from $193 per ton with no international emissions trading to $23 per ton with global trading.[29] COP6 agreed that there would be no quantitative limit on the amount of credit a country could receive from trading, but that domestic action must constitute a significant part of a nation's reduction efforts.[30] With no quantitative limit on trading, any estimate of actual domestic reduction required to comply with the Kyoto Protocol, or of the costs involved, remains problematic — and is moot as long as the United States declines to participate in the Kyoto process.

Conclusion

The precise numerical projections of greenhouse gas emissions (or of proposed reductions) should be viewed as indicative (see **Figure 3**). They are less accurate than they appear, given the potential for revisions in data and the uncertainties of projections. But in assessing the status of U.S. greenhouse gas emissions, the trendline for aggregate greenhouse gas emissions is clear: for the United States, the overall trend is up. None of the reviewed scenarios using assumptions that diminish emissions — low economic growth, putting off retirement of nuclear facilities, accelerated fostering of energy-efficient technologies, the President's voluntary program to reduce greenhouse gas intensity —reverses the upward trend in aggregate greenhouse gas emissions by 2010.[31]

Historical data show that the United States failed to meet its voluntary commitment under the UNFCCC for returning aggregate emissions at the end of the 1990s decade to the 1990 level. Any goal to reduce emissions to or below 1990 levels would require the continuing upward trend to turn down. Even with the potential for sequestration and emissions trading to reduce domestic reduction efforts, a goal to reverse greenhouse gas emissions trends would represent an extraordinary technical and political challenge for U.S. energy and environmental policy.

References

[1] This report does not address the underlying debate over global climate change and the potential role of humans in contributing to it. On the science and policy of global climate change, see CRS Issue Brief IB89005, *Global Climate Change*.

[2] UNFCCC, Article 4, Commitments, sections 2(a) and (b).

[3] On the agreement, see CRS Report RS30692, *Global Climate Change Treaty: The Kyoto Protocol.*

[4] Technically,thenetcarbon-equivalentemissionsofthesixgreenhousegasesforthe5-year period 2008-2012 are not to exceed 5 times 93% of the baseline emissions. Kyoto Protocol, Article 3(1). This is equivalent to the average annual emission load during the 5-year period being 7% below the baseline.

[5] White House, Office of the Press Secretary, "President Bush Discusses Global Climate Change," June 11, 2001.

[6] For documents on the Administration plan, see[http://www.whitehouse.gov/news/releases/2002/02/climatechange.html].

[7] "Bush Emissions Plan Seen Replacing Regs," *Platts Inside Energy* (Feb. 25, 2002), p. 9

[8] Environmental Protection Agency, *Inventory of U.S. Greenhouse Gas Emissionsand Sinks*: 1990-2001 (April 2003), EPA 430R03004.

[9] Although emissions data are typically presented as individual figures for each year, this single number (point estimate) actually represents a range bounded by potential errors arising from assumptions underlying the data.

[10] MMTCE figures combine the variable greenhouse effects of the different gases by calculating and summing their equivalent effects, with carbon dioxide serving as the reference gas. Global warming potential figures have an uncertainty of + or -35%; a recent recalculation is not included in the current figures (see discussion in EPA, *InventoryofU.S. Greenhouse Gas Emissions and Sinks*: 1990-2001 (April 2003), EPA 430R03004. pp. 6-9).

[11] UNFCCC, Article 4, section 1(a) and Article 12, section 1(a).

[12] EPA, *Inventory of U.S. Greenhouse Gas Emissions and Sinks*: 1990 -2001, April 2003, EPA 430R03004, available at [http://yosemite.epa.gov/OAR/globalwarming.nsf/content/ResourceCenterPublicationsGHGEmissionsUSEmissionsInventory2003].

[13] The *Climate Action Report*s, submitted by the United States of America under the United Nations Framework Convention on Climate Change, appeared in October 1994, July 1997, and May 2002; for the third report, see U.S. Department of State, *U.S. Climate Action Report* (May 2002), available at [http://yosemite.epa.gov/oar/globalwarming.nsf/content/ ResourceCenterPublicationsUSClimateActionReport.html].

[14] See "Changes in This Year's Inventory Report," EPA, *Inventory of U.S. Greenhouse Gas Emissions and Sinks*: 1990-2001, pp. Changes-1 -Changes-13.

[15] Following the Administration's practice, discussions in this report of greenhouse gas intensity are based on net emissions per unit of economic activity.

[16] On current federal policies, see CRS Report RL31921, *Climate Change: Federal Laws and Policies Related to Greenhouse Gas Reductions.*

[17] *CAR 2002*, Table 5-2, p. 74; this excludes "Sequestration Removals" and includes "Adjustments."

[18] White House, *Global Climate Change Policy Book,* "Executive Summary," at [http://www.whitehouse.gov/news/releases/2002/02/climatechange.html].

[19] EIA, *Annual Energy Outlook 2003* (Jan. 2003), DOE/EIA-0383(2003).

[20] The impact of economic activity can be seen (**Table 1**) in the fact that gross emissions declined in both 1991 and 2001, in tandem with economic downturns.

[21] DOE, Interlaboratory Working Group, *Scenarios for a Clean Energy Future*, ORNL/CON-476, LBNL-44029, and NREL-TP-620-29379 (Oak Ridge, TN: Oak Ridge NationalLaboratory;Berkeley,CA:LawrenceBerkeleyNationalLaboratory;Golden,CO: NationalRenewableEnergyLaboratory,November2000),at[http://www.ornl.gov/ORNL/Energy_Eff/CEF.htm].

[22] Ibid., p. 1.4.

[23] Ibid., pp. 1.6-1.7. Note that the carbon trading assumption has been effectively mooted since it was not implemented in 2002 as assumed in the study.

[24] See the *Climate Action Report*, Chapter 4; EIA, *Analysis of the Climate Change TechnologyInitiative*,SR/OIAF/99-1,availableat[http://www.eia.doe.gov/oiaf/climate99/climaterpt.html]; and EIA, *Analysis of the Climate Change Technology Initiative: Fiscal Year 2001*, SR/OIAF/2000-01, April 2000, available at [http://www.eia.doe.gov/oiaf/climate/index.html].

[25] Current data suggests that the 2002 carbon intensity baseline may have dropped below 180 MMTCE (see **Table 3**).

[26] "Draft Decision on Implementation of the Kyoto Protocol on Climate Change," adopted in Bonn, Germany, July 23, 2001, footnote to Appendix Z.

[27] Kyoto Protocol, articles 4, 6, and 12; see also CRS Issue Brief IB97057, *Global Climate Change: Market-Based Strategies to Reduce Greenhouse Gases*.

[28] Statement of Janet Yellen, Chair, President's Council of Economic Advisors, House Committee on Commerce, Subcommittee on Energy and Power, March 4, 1998.

[29] For a discussion of the impact of emissions trading on costs, see CRS Report RL30285, *Global Climate Change: Lowering Cost Estimates through Emissions Trading — Some Dynamics and Pitfalls*, by Larry Parker.

[30] "Draft Decision on Implementation of the Kyoto Protocol on Climate Change," adopted in Bonn, Germany, July 23, 2001.

[31] The "advanced" scenario of the "New 5-Lab Study" projects the trend turning downward after 2020.

In: Greenhouse Gases and Their Impact
Editor: Bernard F. Shulk, pp. 157-171

ISBN: 978-1-60021-392-2

Chapter 3

GLOBAL CLIMATE CHANGE: MARKET-BASED STRATEGIES TO REDUCE GREENHOUSE GASES*

Larry Parker

SUMMARY

The possibility that human activities are releasing gases, including carbon dioxide (CO2), at rates that could affect global climate has resulted in proposals for national pro-grams to curtail emissions. An international framework for specific reductions in green-house gases was negotiated at a meeting in Kyoto in December 1997. Concern about costs has encouraged consideration of CO2 reduction proposals that employ market-based mechanisms. The passage in 1990 of a trade-able allowance system for sulfur dioxide (SO2) control in the United States to reduce acid rain provides a precedent for such mechanisms.

The two mechanisms receiving the most attention are a tradeable permit program (similar to the acid rain program) and carbon taxes. Proposed CO2 reduction schemes present large uncertainties in terms of the perceived reduction needs and the potential costs of achieving those reductions. Tradeable permit programs would reduce CO2 emissions to a specific level with the control cost han-dled efficiently, but not ata specific cost level. Carbon taxes would effectively cap marginal control costs at the specific tax level, but the precise level of CO2 reduction achieved would be less certain. Hence, a major policy question is whether one is more concerned about the possible cost of the program and therefore willing to accept some uncertainty about emission reduction in order to have some limits on costs (i.e., carbon taxes) or whether one is more concerned about achieving a specific emission reduction level with costs handled efficiently, but not capped (i.e., tradeable permits).

The specific effects of both a carbon tax and tradeable permit program would depend on the specific levy (carbon tax) or allocation scheme (tradeable permit) chosen, the scope of the program, the timing of the reductions, and the recycling of any revenues.

In addition, many tradeable permit proposals include provisions allowing countries to accumulate permits by reducing emissions in other countries. Two such schemes, joint implementation and the Clean Development Mechanism (CDM), were approved at the Kyoto conference in December 1997.

---

* Excerpted from CRS Report IB97057. Updated November 18, 2004

The climate change issue and CO2 control raise numerous equity issues. In one sense, climate change is a concern about intergenerational equity — the well-being of the current generation versus generations to come. On a global level, the issue also involves the North-South debate. At the domes-tic level, equity questions include the regional distribution of costs under a tradeable permit or carbon tax scheme. For example, an important impact of either a carbon tax based on the carbon content of fossil fuels or a tradeable permit program would be the pressure for fuel shifts away from coal and toward gas. Regions such as fast-growing areas in need of more energy and owners of "all electric" homes, among others, would likely be disproportionately hit by a CO2 control scheme. In addition, people may be affected differently according to income class. These issues, however, have not been sufficiently analyzed at the current time to be sure of how various sectors would be affected.

Most Recent Developments

On November 18, 2004, Russia handed its official Kyoto Protocol ratification papers to the United Nations. With Russia's action, the Kyoto Protocol will become a legally binding treaty on February 16, 2005.

In the 108th Congress, six bills have been introduced to control greenhouse gas emissions. S. 139, introduced by Senators McCain and Lieberman, would reduce and cap emissions of greenhouse gases from electricity generation, transportation, industrial, and commercial sectors. The reductions would be implemented in two phases with an emissions cap in the year 2010 based on affected facilities' 2000 emissions, and a further reduction cap imposed in the year 2016 based on affected facilities' 1990 emissions. The program would be implemented through an expansive allowance trading program that includes cross-sector trading and limited acquisition of allowances from foreign sources. The second bill, S. 366, introduced by Senator Jeffords, is a modified version of the multi-pollutant bill reported out by the Senate Environment and Public Works Committee in the 107th Congress. It would reduce and cap emissions of carbon dioxide from electricity generation at their 1990 levels by the year 2009. Similar to S. 139, the program would be implemented through an allowance trading program. The third and fourth bills, S. 843, introduced by Senator Carper, and H.R. 3093, introduced by Representative Bass are multi-pollutant control bills that include carbon dioxide reductions. S. 843 and H.R. 3093 would cap carbon dioxide emissions from powerplants at their 2006 levels by 2009 and further cap emissions at their 2001 levels by 2013. The fifth bill, H.R. 2042, introduced by Representative Waxman, is also a multi-pollutant control bill and would cap carbon dioxide emissions from powerplants at their 1990 levels by the year 2009. The sixth and most recent bill is H.R. 4067. Introduced March 30, 2004, by Representative Gilchrest, it is modeled on S. 139 but includes only the first phase of S. 139's two-phase reduction program. Like S. 139, H.R. 4067 is a stand-alone greenhouse gas reduction program.

In October 2003, the Senate debated and defeated an amended version of S. 139 (S.Amdt. 2028), the Climate Stewardship Act of 2003. S.Amdt. 2028 would have frozen greenhouse gas emissions from major economic sectors at their 2000 levels, but was defeated on a 43-55 vote. The sponsors of S. 139 have announced their intention to seek another vote on the measure before the end of the 108th Congress.

BACKGROUND AND ANALYSIS

Certain gases emitted as a result of human activities may be affecting global climate. Most concern centers on the possibility that CO2 (primarily from the burning of fossil fuels), along with other gases, could increase global temperatures, with subsequent effects on precipitation patterns and ocean levels that could affect agriculture, energy use, and other human activities.

Status of Global Climate Change Issue and Response

The initial issue of whether the potential for global climate change poses a threat that justifies prompt action to curtail CO2 and other so-called greenhouse gases remains actively debated — both domestically and internationally. (For a review of the technical dimensions of this question, see CRS Issue Brief IB89005, *Global Climate Change*.) Some view the risks as sufficiently grave and urgent to justify immediate action. Others are uncertain of the risks but believe that selected policies to reduce emissions can be justified for other reasons and would provide insurance if the risks were borne out; these other reasons include improved energy efficiency, reduced reliance on imported oil, and increased revenues. Still others caution that actions to reduce CO2 and other greenhouse gases could disrupt the nation's economy and should not be undertaken unless further scientific evidence of risks becomes available.

Despite the uncertainties, however, scientists and policymakers have increasingly adopted the view that human activities are releasing greenhouse gases at rates that could affect global climate. As a result, initiatives are underway to address the issue, resulting in proposals for national and international programs to curtail emissions.

An agreement on a United Nations Framework Convention on Climate Change (UNFCCC) was on the agenda at the U.N. Conference on Economic Development in Rio de Janeiro in June 1992. The United States was an early signatory to the agreement, which was approved by the Senate October 7, 1992. In April 1993, President Clinton directed the federal government to craft a plan that would stabilize U.S. greenhouse gas emissions at 1990 levels by the year 2000. However, in 2000, the United States did not meet its voluntary commitment at Rio to stabilize greenhouse gas emissions at 1990 levels. Indeed, it is unclear when U.S. carbon emissions may stabilize. The 2002 *Climate Action Report* by the current Bush Administration estimates U.S. carbon emissions in the year 2010 will be 34% above their 1990 levels. (For more on U.S. domestic climate change policy since Rio, see CRS Report RL30024, *Global Climate Change Policy: Cost, Competitiveness, and Comprehensiveness*.)

Meanwhile, the United States and other signatories to the Climate Change Convention met in December 1997 in Kyoto, Japan, to conclude negotiations on a binding protocol for specific provisions to reduce greenhouse gas emissionsby developed countries, including the United States. The final protocol agreed to at Kyoto requires the United States to reduce emissions of six greenhouse gases (CO2, methane, nitrous oxide, hydrofluorocarbons, perfluorocarbons, sulfur hexafluoride) by 7% on average from 1990 levels over the period 2008-2012. In contrast, undeveloped countries are not required to make reductions under the Protocol. In November 1998, the parties met in Buenos Aires to develop work plans for specific elements of the Kyoto Protocol, including the trading of emission reductions between developed countries (joint implementation) and between developed and developing countries (the Clean Development Mechanism). The parties decided that these work plans should be completed by the year 2000. The November 1999 meeting in Bonn postponed decisions about emissions trading until the November 2000 meeting at the Hague. The meeting at The Hague failed to arrive at agreement on emissions trading, and scheduled further negotiations for May 2001. However, in March 2001, the current Bush Administration announced that it was formally abandoning the emission targets set under Kyoto. (For more on the U.S. reduction requirement under Kyoto, see CRS Report 98-235 ENR, *Global Climate Change: U.S. Greenhouse Gas Emissions — Status, Trends, and Projections.*)

This decision bythe Bush Administration has not deterred the international community. In July 2001, the Sixth Conference of Parties to the Framework Convention on Climate Change agreed to a draft decision on implementing the Kyoto Protocol. With respect to flexible implementation mechanisms, the Parties agreed to exclude nuclear power as a possible non-carbon alternative under the Clean Development Mechanism (CDM) and joint implementation program. The Parties also reiterated that use of flexible mechanisms shall be supplemental to domestic efforts. In November 2001, the Seventh Conference of Parties to the Framework Convention on Climate Change concluded negotiations on implementation of the Kyoto Protocol. With respect to flexible implementation mechanisms, the Parties outlined the institutions that would oversee the flexible implementation mechanisms contained in the Protocol. In October 2002, the Eighth Conference of Parties to the Framework Convention on Climate Change met in New Delhi (October 23 - November 1). The conference produced agreement on the procedures governing the Clean Development Mechanism. In December 2003, the Ninth Conference of Parties met at Milan and produced agreement on rules governing the use of sinks for joint implementation and CDM projects. The United States was a non-participant to most of what took place at these post-March 2001 conferences.

The Kyoto Protocol entered a new phase on November 18, 2004, when Russia handed its official Kyoto Protocol ratification papers to the United Nations. With Russia's action, the Kyoto Protocol will become a legally binding treaty on February 16, 2005. UN Secretary General Kofi Annan described Russia's accession to the treaty as a "historic step forward in the world's efforts to combat a truly global threat."

Estimating Cost Impacts of Controls

Estimates of costs to reduce CO2 emissions vary greatly and focus attention on an estimator's basic beliefs about the problem and the future, rather than on simple, technical differences in economic assumptions. (See CRS Report 98-738, *Global Climate Change: Three Policy Perspectives*. It identifies three "lenses" through which people can view the global climate change issues, and their influence on cost analysis.) These are summarized in Table 1. None of these perspectives is inherently more "right" or "correct" than another; rather, they overlap and to varying degrees complement and conflict with each other. People hold to each of the lenses to some degree.

However, the differing perspectives lead to very different cost estimates. Figure 1 below shows a scatter-plot by World Resources Institute (WRI) of the predicted impacts from 162 estimates from 16 different economic models on the U.S. economy from a CO2 abatement program. Although the size of the proposed CO2 reduction and the time allowed to achieve it (not explicitly modeled in the WRI report) are critical factors in determining the costs and benefits of any reduction program, WRI found that underlying modeling assumptions not related to policy decisions explained a significant amount of the difference in the estimates. Consistent with a "technological" view of the problem, models that assumed technological development of non-carbon substitutes for current fossil fuel use, along with increased energy and product substitutions, had significantly less cost than models that assumed such advancements would not occur in a timely fashion. For example, a study by the American Council for an Energy-Efficient Economy (ACEEE) argues that carbon emissions could fall 10% below 1990 levels by 2010 with a net economic savings of $58 billion along with 800,000 new jobs. Such savings are assumed to come from new technology and market mechanisms to encourage cost-effective implementation strategies. Such a position presumes that technologies are available now, or will be very shortly, that can achieve these reductions cost-effectively.

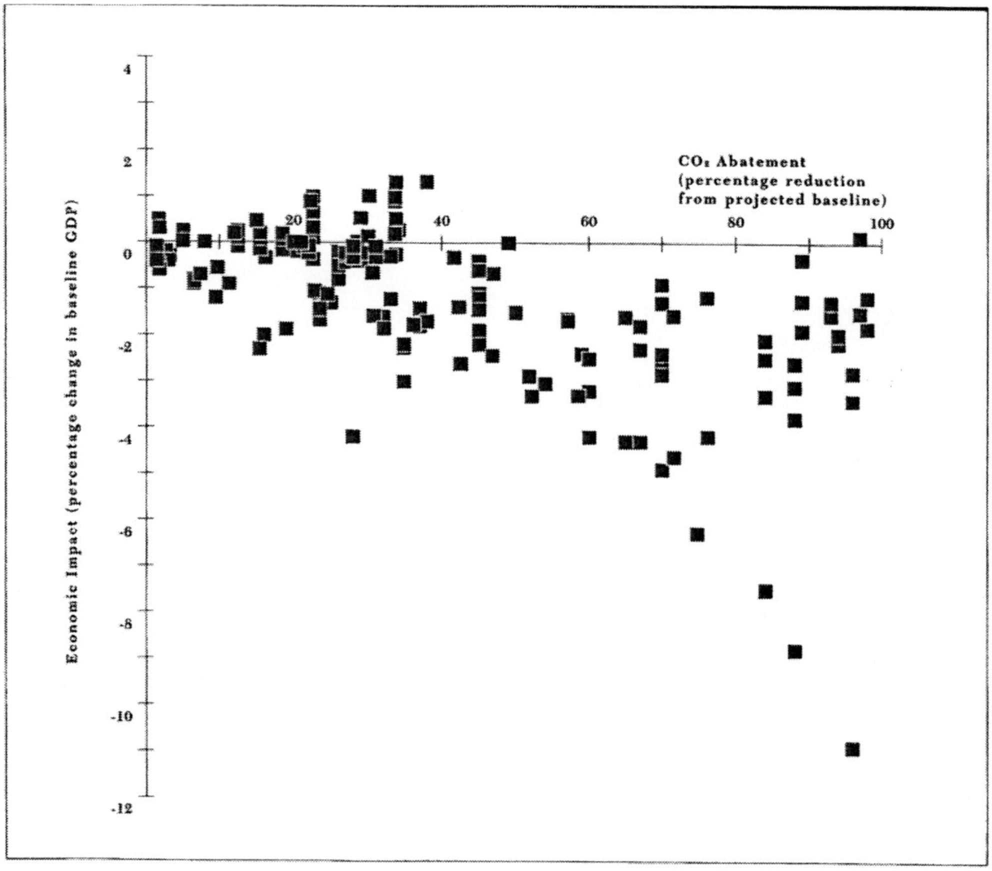

Figure 1. The predicted impacts of carbon abatement on the U.S. economy (162 estimates from 16 models)

Likewise, consistent with an "ecological" perspective, models that included the benefits of air pollution damages and climate change damages averted by the CO2 reduction estimated considerably less cost to the economy than models that did not include such benefits. The WRI report suggests that the cost profile of a CO2 reduction program changes substantially if one includes the benefits of air pollution and climate change effect averted by controlling

CO_2. The Clinton Administration's 1998 analysis of costs to comply with Kyoto estimates benefits from controlling ancillary pollutants (SO_2, NO_x, and fine particulates) at between \$1.8 and \$10.6 billion annually.

Consistent with an "economic" perspective, models that included policy approaches that encouraged efficient economic responses to CO_2 reductions, included joint implementation schemes, and involved efficient recycling of any revenues from control strategies significantly reduced costs over model runs that did not include such policy options. Like the technology perspective, economically efficient solutions assume that the program is implemented in such a way to permit the economy sufficient time to absorb the new price signals with minimal short-term constraints.

The uncertainty about the risk of climate change and the critical impact of assumptions about the nature of the problem effectively preclude predictions of the ultimate costs of reducing greenhouse gases. As a result, attention has focused on how to minimize costs by selecting the most economically efficient strategies to reduce CO_2 emissions. Traditionally, air pollution control programs have relied on various "command and control" regulatory approaches, including ambient quality and technology-based standards. But increasingly, economic efficiency concerns have been directed toward supplementing regulatory control with market-based mechanisms, including pollution taxes and tradeable permits.

The tradeable allowance system for SO_2 control in the acid rain program enacted in 1990 represents a significant step in this evolution of economic mechanisms. Acceptance of this system has led to calls for use of a similar system with other pollutants, including CO_2. Current bills proposing a tradeable permit-type system to begin controlling CO_2 emissions are discussed in a later section.

Table 1. Influence of Climate Change Perspectives on Policy Parameters

Approach	Seriousness of problem	Risk in developing mitigation program	Costs
Technology	Is agnostic on the merits of the problem. The focus is on developing new technology that can be justified from multiple criteria, including economic, environmental and social perspectives.	Believes any reduction program should be designed to maximize opportunities for new technology. Risk lies in not developing technology by the appropriate time. Focus on research, development, and demonstration; and on removing barriers to commercialization of new technology.	Viewed from the bottom up. Tends to see significant energy inefficiencies in the current economic system that currently (or projected) available technologies can eliminate at little or no overall cost to the economy.

Table 1(continued)

Approach	Seriousness of problem	Risk in developing mitigation program	Costs
Economic	Understands issue in terms of quantifiable cost-benefit analysis. Generally assumes the status quo is the baseline from which costs and benefits are measured. Unquantifiable uncertainty tends to be ignored.	Believes that economic costs should be examined against economic benefits in determining any specific reduction program. Risk lies in imposing costs in excess of benefits. Any chosen reduction goal should be implemented through economic measures such as tradeable permits or emission taxes.	Viewed from the top down. Tends to see a gradual improvement in energy efficiency in the economy, but significant costs (quantified in terms of GDP loss) resulting from global climate change control programs. Typical loss estimates range from 1-2% of GDP.
Ecological	Understands issues in terms of their potential threat to basic values, including ecological viability and the well-being of future generations. Such values reflect ecological and ethical considerations; adherents see attempts to convert them into commodities to be bought and sold as trivializing the issue.	Rather than economic costs and benefits or technological opportunity, effective protection of the planet's ecosystems should be the primary criterion in determining the specifics of any reduction program. Focus of program should be on altering values and broadening consumer choices.	Views costs from an ethical perspective in terms of the ecological values that global climate change threatens. Believes that values such as intergenerational equity should not be considered commodities to be bought and sold. Costs are defined broadly to include aesthetic and environmental values that economic analysis cannot readily quantify and monetize.

Market-Based Mechanisms for Reducing Greenhouse Gases

Proposals to use market mechanisms to implement greenhouse gas emission reductions have revolved around three approaches: tradeable permits (as "allowances" and as "credits"), carbon taxes, and joint implementation. The protocol negotiated at Kyoto contains articles on emissions trading and joint implementation. These provisions were strongly supported by the Clinton Administration. In addition, some European countries have implemented or are considering carbon taxes to bring about greenhouse gas reductions in their countries.

Tradeable Permits (Allowances)

A model for a tradeable permit approach is the SO2 allowance program to reduce acid rain contained in Title IV of the 1990 Clean Air Act Amendments. The Title IV program is based on two premises. First, a set amount of SO2 emitted by human activities can be assimilated by the ecological system without undue harm. Thus the goal of the program is to put a ceiling, or cap, on the total emissions of SO2 rather than limit ambient concentrations. Second, a market in pollution rights between polluters is the most cost-effective means of achieving a given reduction. This market in pollution rights (or allowances, each of which is equal to one ton of SO2) is designed so that owners of allowances can trade those allowances with other emitters who need them or retain (bank) them for future use or sale. Initially, most allowances were allocated by the federal government to utilities according to statutory formulas related to a given facility's historic fuel use and emissions; other allowances have been reserved by the government for periodic auctions to ensure the liquidity of the market.

Conceptually, a CO2 tradeable permit program could work similarly. Some number of CO2 allowances could be allocated, and a market in the allowances would permit emitters to use, sell, buy, or bank them. However, significant differences exist between acid rain and possible global warming that may affect the appropriateness of a Title IV-type response to CO2 control. For example, the acid rain program involves up to 3,000 new and existing electric generating facilities that contribute two-thirds of the country's SO2 and one-third of its nitrogen oxide (NOx) emissions (the two primary precursors of acid rain). This concentration of sources makes the logistics of allowance trading administratively manageable and enforceable. However, CO2 emissions are not so concentrated. Although over 95% of the CO2 generated comes from fossil fuel combustion, only about 33% comes from electricity generation. Transportation accounts for about 33%, direct residential and commercial use about 12%, and direct industrial use about 20%. Thus, small dispersed sources in transportation, residential/commercial, and the industrial sectors are far more important in controlling CO2 emissions than they are in controlling SO2 emissions. This creates significant administrative and enforcement problems for a tradeable permit program if it attempts to be comprehensive.

These concerns multiply as the global nature of the climate change issue is considered, along with other potential greenhouse gases. Article 3 of the protocol negotiated at Kyoto emphasizes that any international emissions trading should be supplemental to a country's domestic efforts, not a substitute for them.

Tradeable Permits (Credits)

As noted above, a tradeable allowance involves future emissions. An allowance is a limited authorization to emit a ton of pollutant; allowances are allocated to an emitting facility under an applicable emission limitation at the beginning of a year. The facility decides whether to use, trade, or bank those allowances, depending on its emissions strategy. Then, atthe end of the year, the agency compares anemittingfacility's actual emissions with its available allowances to determine compliance.

A different approach to creating a tradeable permit program is to use credits instead of allowances. A credit is created when a facility actually emits a pollutant at less than its allowable limit as defined in by the program. An example of this type of program is the Environmental Protection Agency's (EPA) "Emission Reduction Credit program" (ERC) under the Clean Air Act. Under the ERC program, EPA requires that any credit created under a state program implementing emissions trading be "surplus, enforceable (by the state), permanent, and quantifiable." Thus, a state must certify the creation of the credit, unlike an allowance program, where allocation is dictated by a statutory or regulatory formula. Any CO2 reduction credit program could build on EPA's and states' experience with the current emission reduction credit program.

The primary advantage of a credit program over an allowance program is that it does not discriminate against new sources. Allowance programs tend to allocate their allowances based on some historic baseline year. Those sources included in the baseline get their allowances free. Those future sources not included in the baseline have to pay either the older, existing sources to obtain allowances or to buy allowances at auction. With a credit program, sulfur credits can be created by any source, as the baseline is dictated by the emissions cap and yearly production, not a historical year. The disadvantage of such a system is that facility planning is very difficult as operators do not know precisely what their permissible limit will be from year to year.

Carbon/CO2 Emissions Tax

An alternative market-based mechanism to the tradeable permit system is carbon taxes — generally conceived as a levy on natural gas, petroleum, and coal according to their carbon content, in the approximate ratio of 0.6 to 0.8 to 1, respectively. In the view of most economists, the most efficient approach to controlling CO2 emissions would be a carbon tax. With the complexity of multiple pollutants and millions of emitters involved in controlling CO2, the advantages of a tax are self-evident. Imposed on an input basis, administrative burdens such as stack monitoring to determine compliance would be reduced. Also, a carbon tax would have the broad effect across the economy that some feel is necessary to achieve long-term reductions in emissions.

However, in other ways, a tax system merely changes the forum rather than the substance of the policy debate. Because paying an emissions tax becomes an alternative to controlling emissions, the debate over the amount of reductions necessarily becomes a debate over the tax level imposed. Those wanting large reductions quickly would want a high tax imposed over a short period of time. Those more concerned with the potential economic burden of a carbon tax would want a low tax imposed at a later time with possible exceptions for various events. Emissions taxes would remain basically an implementation strategy; policy determinations such as tax levels would require political/regulatory decisions. In addition, a tax system would raise revenues. Indeed, one argument for — or against — such a system would be that it is a tax that would raise revenues. The disposition of these revenues would significantly affect the economic and distributional impacts of the tax.

Other tax schemes to address global climate change are also possible. For example, the European Community (EC) has discussed periodically a hybrid carbon tax/energy tax to begin addressing CO2 emissions. Fifty percent of the tax would be imposed on energy production (including nuclear power) except renewables; 50% of the tax would be based on carbon emissions. Some European countries have modified their energy taxation to fit the model discussed by the EC.

Currently, five European countries have carbon-based taxes. Finland imposed the first CO2 tax in 1990 and modified it in 1994. The Finnish tax has two components: (1) a basic tax component to meet fiscal needs and (2) a combined energy/CO2 tax component. For coal, peat, and natural gas, there is no fiscal component. The Netherlands also introduced a CO2 tax in 1990, modified in 1992 to fit the EC model; however, the tax does not vary according to fuel type and energy use. It does include tax relief from the energy component of the tax for energy-intensive industries. Sweden introduced a CO2 tax in 1991 on all fossil fuels, unless they are used in electricity production. In 1993, the tax scheme was modified to reduce its burden on industry. Denmark introduced a CO2 tax in 1992 that covers fuel oil, gas, coal, and electricity (gasoline is taxed separately). Taxes paid by industry are completely reimbursed to the sector. Norway introduced a CO2 tax in 1991 on oil and natural gas and extended it to some coal and coke use in 1992. However, there are many exemptions and the tax rate is not differentiated according to the carbon content of the fuels.

International Market-Oriented Mechanisms

Joint Implementation (JI) and the Clean Development Mechanism (CDM) are attempts to expand the availability of cost-effective CO2 reductions into the international sphere through flexible, market-oriented mechanisms. Basically, these mechanisms allow a developed country needing to make CO2 reductions to meet its obligations under an international treaty to obtain reduction credits by financing emission reductions in another country: either another developed country in the case of JI, or a developing country in the case of CDM. During the 1990s, promoting flexible market-oriented mechanisms was a keystone of U.S. climate change policy; they were subject to considerable debate at the Conference of Parties (COP) meetings in Berlin. These discussions resulted in agreement to implement JI in a pilot phase (renamed Activities Jointly Implemented — AJI).

After much negotiation, the protocol agreed to at Kyoto contains provisions on flexible market-oriented mechanisms that generally followed the guidelines set up at Berlin. Article 6 provides for JI between developed countries with emission reduction requirements. Because developing countries have no emission requirements to meet (unlike developed countries), article 12 of the protocol set up a Clean Development Mechanism (CDM) to promote sustainable development in them while providing emission reduction opportunities for developed countries. Participation is voluntary; benefits must be real, measurable, and long-term; reductions must be in addition to any normal activity. Operated under supervision of the COP, reductions achieved between 2000 and 2008 may be used to offset commitments in the 2008-2012 time period.

The advantage of flexible, market-oriented mechanisms for developed countries is that they widen the options available to obtain necessary credits under any reduction program. This translates into lower costs to those countries, compared with their own domestic reduction activities. For the developed country, particularly where it does not have the resources to control emissions or protect sequestration areas, reductions or protection would occur more quickly than would otherwise be possible.

However, the disadvantages are also significant. A developed country may have to rely on another sovereign government to ensure compliance with part of its international commitment. Governments change, and policies change. If a new government chose to remove or shut down a pollution control device, the developed country might have little recourse but to look elsewhere for its necessary reduction. Particularly with sequestration projects that involve marketable commodities, such as trees, enforcement could be quite difficult. A tree's value as cooking or heating firewood for natives could easily exceed its value as a carbon sequester. In the long-run, the enthusiasm with which a developing country may enforce agreements with respect to CDM projects is unclear.

Indeed, developing countries could have significant economic incentives to abrogate CDM projects, particularly if they are viewed as constraining necessary development, or locking up a natural resource that the country would like to exploit. This incentive is further encouraged if the CDM project is perceived as a developed country's project. The term "economic imperialism" has already been applied to CDM projects by some opponents.

Issues

Cost-Effectiveness: Price versus Quantity

Proposed CO2 reduction schemes present large uncertainties in terms of the perceived reduction needs and the potential costs of achieving those reductions. In one sense, preference for a carbon tax or tradeable permit system depends on how one views the uncertainty of costs involved and benefits to be received. For those confident that achieving a specific level of CO2 reduction will yield very significant benefits — enough so that even the potentially very high end of the marginal cost curve does not bother them — then a tradeable permit program may be most appropriate. CO2 emissions would be reduced to a specific level, and in the case of a tradeable permit program, the cost involved would be handled efficiently, but not controlled at a specific cost level. This efficiency occurs because control efforts are concentrated at the lowest-cost emission sources through the trading of permits.

However, if one is more uncertain about the benefits of a specific level of reduction — particularly with the potential downside risk of substantial control cost to the economy — then a carbon tax may be most appropriate. In this approach, the level of the tax effectively caps the marginal control costs that affected activities would have to pay under the reduction scheme, but the precise level of CO2 achieved is less certain. Emitters of CO2 would spend money controlling CO2 emissions up to the level of the tax. However, since the marginal cost of control among millions of emitters is not well known, the overall effect of a given tax level on CO2 emission cannot be accurately forecasted. Hence, a major policy question is whether one is more concerned about the possible economic cost of the program and therefore willing to accept some uncertainty about the amount of reduction received (i.e., carbon taxes) or whether one is more concerned about achieving a specific emission reduction level with costs handled efficiently, but not capped (i.e., tradeable permits).

A proposal was floated by the Clinton Administration for a tradeable permit program with a ceiling on the price of permits. If permit prices rose above a certain level, the government would have intervened to control costs by selling more permits at a specific price. In essence, this would have given the permit program the character of a carbon tax by controlling costs through a price "safety valve," while allowing quantity to increase to any level necessary to prevent price increases. Not surprisingly, environmental groups interested in protecting the emission limitations of any global climate change program attacked the idea as a "target-busting escape clause." Industry groups suggested that such a tradeable permit program amounts to a tax. For a discussion of other "safety valve" proposals, see CRS Report RS21067, *Global Climate Change: Controlling CO2 Emissions — Cost-limiting Safety Valves*.

Comprehensiveness

As suggested earlier, carbon emissions are ubiquitous. Much of the emissions come from the direct combustion of fossil fuels from small, dispersed sources such as automobiles, homes, and commercial establishments. For example, the 12% of U.S. emissions from the residential/commercial sector come from such thingsas space heating/cooling (9.3%, oil and natural gas), water heating (1.5%, mostly natural gas), and appliances (1.2%, mostly natural gas). If one adds to these dispersed sources the 33% of emissions that come from direct combustion from automobiles (13.9%), trucks (11.2%), airplanes (4.5%), ships (1.8%), pipelines (0.6%), and railroads (0.8%), the number of individual sources runs into the millions; very small sources contribute almost half the emissions.

Assuming a carbon tax is assessed on an input basis (i.e., on the carbon content of the fuel), then the number of sources is largely irrelevant — the sources would get the correct price signal from the increased cost of their fuel. This is one of the primary strengths of the carbon tax scheme — it can be very comprehensive and potentially induce the necessary changes in individual as well as corporate behavior that could substantially reduce dependence on carbon emitting energy sources. In this sense, a carbon tax is not just a band-aid to reduce CO2 emissions, but a program to reduce carbon intensiveness in the economy and in individual lifestyles.

For a tradeable permit program, the numbers of sources can represent a substantial administrative and enforcement problem. One approach to making the situation more manageable would be to limit the scope of the trading system to domestic implementation strategies. As noted above, international emission trading is termed "supplemental" under the consolidated negotiating text. Likewise, the scope could be limited further by focusing the trading program on the electric utility sector. Another approach could be to limit the size of the source included in the trading program. Others could "opt-in," but their participation would be voluntary. Thus, direct combustion of fossil fuels in the residential, commercial, and industrial sectors (e.g., natural gas, home heating oil) would be indirectly encouraged by the program and use of CO2 emitting electricity (particularly coal-fired electricity) discouraged. The transportation sector would be little affected (unless it chose to be).

Economic Impact

Obviously, the economic impact of either a tradeable permit program or a carbon tax depends on the level of reductions desired and the timing of those reductions. Most of the studies on the economic impact of CO2 control programs have focused primarily on carbon taxes. This is not surprising as carbon taxes are easier to model than a tradeable permit program. However, the uncertainty involved in these analyses is quite large; further work is necessary to reduce the current range of estimates.

Economic assumptions that influence cost estimates include (1) carbon emissions growth assumptions in the absence of legislation, (2) responsiveness of the economy to the carbon tax in terms of increased energy efficiency, and (3) type of model employed. Uncertainty is compounded when attempts are made to estimate GDP effects of carbon taxes. Very small differences in GDP estimation techniques can result in large differences in projected impacts (particularly over the long term). Preliminary evidence indicates that the adverse effects of a carbon tax can be reduced if the proceeds from that tax are "recycled" either to offset certain existing taxes or fund investment incentives to encourage economic growth (particularly through greater capital formation). Thus, the impact of a carbon tax on the economy would depend to some degree on how the government disposed of generated revenues. However, considerably more work is needed to define the economic consequences of a specific proposal to recycle revenues before much confidence can be put into the results. Of course, if one has a technological or ecological orientation, the assumptions resulting from those orientations can outweigh the economic assumptions discussed here.

The extent that economic analysis of carbon tax programs provides insight for a tradeable permit program depends partially on the scope of the program, the options included, and the monitoring and transaction costs. If the government chose to sell its allowances at auction, rather than giving them away (as is typical), the government would have revenue like a carbon tax to recycle or readdress perceived distortions in the current tax code. In June 2000, the Congressional Budget Office released a study on the distributional effects of carbon trading programs. It concludes that if the government gave away carbon allowances to U.S. firms (as is typical for trading programs), the effects would be regressive on households. If the allowances were sold at auction, the distributional effects would depend on the ultimate disposition of the revenue received from the sale. However, the carbon tax analysis does suggest that the price of a permit (and any revenues from the sale thereof) would be difficult to estimate with any precision at the current time.

The specific effects of both a carbon tax and tradeable permit program would depend on the specific levy (carbon tax) or allocation scheme (tradeable permit) chosen. Experience with both tax code revisions and the allocation scheme under the acid rain title suggests that regional, state, and sector-specific concerns could receive special treatment in these decisions. In addition, for a carbon tax, the allocation of revenue received could also be influenced by such concerns.

Equity

The climate change issue and CO2 control raise numerous equity issues. In one sense, the concern about climate change is a concern about intergenerational equity — the well-being of the current generation versus generations to come. On a global level, the issue also involves the North-South debate. Some industrialized Northern countries suggest that the lesser-developed Southern countries refrain from certain activities (such as clearing rain forests) that Southern countries feel are important for their economic growth. Southern countries often suggest that the Northern countries change their current "unsustainable" growth practices and assist the South in sustainable development. Some supporters of tradeablepermits have suggested that internationalization of the permit program could allow the wealthy countries to fund CO2-reducing activities (preserving forests, improving efficiency, etc.) as a means of achieving cost-effective reductions and assisting developing countries (i.e., CDM projects). However, as noted above, monitoring the long-term efficacy of CDM projects raises administrative issues. Some carbon tax proponents have suggested that a portion of collected revenue could be set aside for assisting developing countries. Percentages to be set aside and more generally the political acceptability of such a proposal are unclear.

Other equity questions include the regional distribution of costs under a tradeable permit or carbon tax scheme. For example, an important impact of either a carbon tax based on the carbon content of fossil fuels or a tradeable permit program would be the pressure for fuel shifts away from coal and toward gas. Other regions, such as fast growing areas in need of more energy and owners of "all electric" homes, among others, would likely be disproportionately hit by a CO2 control scheme. In addition, people may be affected differently according to income class. These issues have not been sufficiently analyzed at the current time to draw firm conclusions.

Legislation in the 108th Congress

In the 108th Congress, six bills have been introduced to control CO2 emissions. S. 139, introduced by Senators McCain and Lieberman, would reduce and cap emissions of carbon dioxide from electricity generation, transportation, industrial, and commercial sectors — sectors that account for about 85% of U.S. greenhouse gas emissions. The reductions would be implemented in two phases with an emissions cap in the year 2010 based on affected facilities' 2000 emissions, and a further reduction cap imposed in the year 2016 based on affected facilities' 1990 emissions. The program would be implemented through an expansive allowance trading program that includes cross-sector trading and limited acquisition of allowances from foreign sources. For a comparison of S. 139 with other Senate bills to reduce greenhouse gases, see CRS Report RS21581, *Climate Change: Senate Proposals to Reduce Greenhouse Gas Emissions*. For a summary and analysis of S. 139, see CRS Report RS21637, *Climate Change: Summary and Analysis of the "Climate Stewardship Act" (S. 139/S.Amdt. 2028, and H.R. 4067)*.

The second bill, S. 366, introduced by Senator Jeffords, is a modified version of the multi-pollutant bill (S. 556) reported out by the Senate Environment and Public Works Committee in the 107th Congress. Placing emission caps on nitrogen oxides, sulfur dioxide, and carbon dioxide, S. 366 would reduce and cap emissions of carbon dioxide from electricity generation at their 1990 levels by the year 2009. Similar to S. 139, the program would be implemented through an allowance trading program. In addition to these emission caps, S. 366 would place facility-specific emission limitations on mercury.

The third and fourth bills, S. 843, introduced by Senator Carper, and H.R. 3093, introduced by Representative Bass, are multi-pollutant control bills similar to S. 3135 introduced in the 107th Congress. Placing emission caps on nitrogen oxides, sulfur dioxide, mercury, and carbon dioxide, S. 843 and H.R. 3093 would cap carbon dioxide emissions from powerplants at their 2006 levels by 2009 and further cap emissions at their 2001 levels by 2013. Similar to the other two control bills, S. 843 and H.R. 3093 would be implemented through an allowance trading program.

The fifth bill, H.R. 2042, introduced by Representative Waxman, is a multi-pollutant control bill similar to H.R. 1256 introduced in the 107th Congress. Placing emission caps on nitrogen oxides, sulfur dioxide, mercury, and carbon dioxide, H.R. 2042 would cap carbon dioxide emissions from powerplants at their 1990 levels by the year 2009. Implementation strategies are to be determined byEPA with market mechanisms explicitly permitted (except for mercury). For a further discussion of multi-pollutant legislation introduced in the 108th Congress, see CRS Report RL31779, *Air Quality: Multi-Pollutant Legislation in the 108th Congress*.

The sixth and most recent bill is H.R. 4067. Introduced March 30, 2004, by Representative Gilchrest, it is modeled on S. 139 but includes only the first phase of S. 139's two-phase reduction program. Like S. 139, H.R. 4067 is a stand-alone greenhouse gas reduction program.

In October 2003, an amended version of S. 139 (S.Amdt. 2028) was debated and defeated on the Senate floor. The vote was 43-55.

Other Proposals

United States and International Activities

In March 2001, the Bush Administration announced that the Kyoto Protocol was "dead" as far as it was concerned. In rejecting the Kyoto Protocol as unfair to the United States, EPA Administrator Whitman emphasized the Administration's desire to work constructively with the EC to develop technologies, market-based incentives, and other innovative approaches to global climate change. However, the Administration has yet to announce or outline any policy alternatives as a basis for international discussions.

Administration Domestic Initiatives

In late September 2000, presidential candidate George W. Bush proposed a national energy plan that would include requiring utilities to reduce their carbon dioxide emission over a "reasonable" time frame in a manner similar to the current market-based acid rain reduction program. Few specifics, such as reduction targets or schedule, were included in the plan. In March 2001, the Bush Administration reversed this position, stating that it would not seek legislation to reduce CO2 emissions. In making the reversal, the Administration cited a Department of Energy (DOE) study indicating that energy costs would increase if controls were put on CO2 emissions.

In February 2002, the Administration initiated a new voluntary greenhouse gas program, similar to ones introduced by the earlier Bush and Clinton Administrations. Developed in response to the U.S. ratification of the 1992 United Nations Framework Convention on Climate Change (UNFCCC), these previous plans projected U.S. compliance, or near compliance, with the UNFCCC goal of stabilizing greenhouse gas emissions at their 1990 levels by the year 2000 through voluntary measures. The new proposal introduced by the Bush Administration did not make that claim, only projecting a 100 million metric ton reduction in emissions from what would occur otherwise in the year 2012. The plan focuses on improving the carbon efficiency of the economy, reducing current emissions of 183 metric tons per million dollars of GDP to 151 metric tons per million dollars of GDP in 2010. It proposes several voluntary initiatives, along with increased spending and tax incentives, to achieve this goal. However, the Administration projects that three-quarters of this reduction would be achieved through current efforts underway, not by the new initiatives.

Based on the Administration's estimates, the initiative will result in U.S. greenhouse gas emissions being 28% above 1990 levels in the year 2010, a 4.5% reduction over a business-as-usual baseline.

INDEX

C

D

E

H

I

J

K

L

M

N

R

S

T

U